Media, Journalism and Disaster Communities

Jamie Matthews • Einar Thorsen
Editors

Media, Journalism and Disaster Communities

palgrave
macmillan

Editors
Jamie Matthews
Bournemouth University
Bournemouth, UK

Einar Thorsen
Bournemouth University
Bournemouth, UK

ISBN 978-3-030-33711-7 ISBN 978-3-030-33712-4 (eBook)
https://doi.org/10.1007/978-3-030-33712-4

Cover design: eStudioCalamar
Cover illustration: Craig Stennett / Alamy Stock Photo

This Palgrave Macmillan imprint is published by the registered company Springer Nature Switzerland AG.
The registered company address is: Gewerbestrasse 11, 6330 Cham, Switzerland

Contents

Notes on Contributors

Aida Al-Kaisy is a media reform advisor and has worked extensively on media development projects across the Middle East and North Africa (MENA) region including in Iraq, Palestine and Tunisia. She is working on projects focusing on issues related to youth engagement in media, media in conflict, social cohesion and the media and the development of independent media platforms in MENA. Her PhD focus is on the performance of the media in conflict, using Iraq as a case study, and she teaches at the School of Oriental and African Studies, University of London, on a part-time basis.

Kurt Barling is Professor of Journalism at Middlesex University London. He is an award-winning investigative journalist who was BBC special correspondent in London from 2001 to 2014 and reported extensively on the Lakanal fire disaster. Barling is author and editor of four books including the international bestsellers *Darkness over Germany* (2016), *The R Word: Racism* (2015) and *Abu Hamza: Guilty* (2014). He previously taught International Relations at the London School of Economics and Political Science (LSE).

Lisa Brooten is an associate professor at Southern Illinois University Carbondale and conducts research in Myanmar, Thailand, the Philippines and the United States. She has published widely on media reform, media representations of marginalised groups, media and human rights, social movement media and various forms of media activism.

Mathew Charles is a freelance foreign correspondent based in Bogotá, working for the BBC and *The Telegraph*. He has experience of both reporting and researching organised crime, urban violence and conflict across Latin America. He has a doctorate from Cardiff University, where he is a visiting research fellow in the School of Journalism, Media and Culture.

Florencia Enghel is an associate professor at Malmö University's School of Arts and Communication and studies the links between communicative practices, media uses and processes of development cooperation and social change. Her work has appeared in *Communication Theory*, *Media, Culture & Society* and *Journal of International Communication*, among others. She is the co-editor of *Communication in International Development: Doing Good or Looking Good?* (2018).

Jacqui Ewart was a journalist and media manager for more than a decade. She is Professor of Journalism and Media Studies in the School of Humanities, Languages and Social Science at Griffith University. Her research focuses on disaster communication, news media representations of cultural and religious diversity, talk radio audiences and regional news media. Ewart has authored, co-authored and edited five books and has held several nationally competitive funded grants.

Kyle J. Holody is an associate professor in the Department of Communication, Media and Culture at Coastal Carolina University. He holds a PhD in Media and Communication from Bowling Green State University. His research focuses on news coverage of controversial health and safety concerns, including mass shootings, abortion, disproportionate treatment based on race and ethnicity and disordered eating.

Victoria Jack, PhD, is a consultant at the UN Migration Agency and an adjunct lecturer at University of New South Wales. Her research focuses on the role of community engagement in humanitarian and development work, especially in contexts of migration.

Jamie Matthews, PhD, is Senior Lecturer in Communication and Media at Bournemouth University. His research interests include international communication, journalism studies, with a particular interest in disaster journalism, media framing of crises and conflict and its intersection with public opinion dynamics.

Dariya Orlova is a media researcher and senior lecturer at the Mohyla School of Journalism, Kyiv, Ukraine. She holds a PhD in Mass Communications from the Autonomous University of Barcelona. She was a visiting assistant professor at CREEES, Stanford University, in 2016. Her research focuses on developments in Ukrainian journalism and media.

Mervi Pantti is Professor of Media and Communication Studies at the Faculty of Social Sciences, University of Helsinki. Her research is concerned with the emotional dimension of mediated communication, crisis and disaster journalism, and humanitarian communication. She has published close to 100 articles in academic journals and edited books. She is co-author of *Disasters and the Media* (with Karin Wahl-Jorgensen and Simon Cottle, 2012) and editor of *Media and The Ukraine Crisis: Hybrid Media Practices and Narratives of Conflict* (2016). Currently, she serves as associate editor of *Journalism Studies.*

Juliet Pinto, PhD, is an associate professor at The Pennsylvania State University's Donald P. Bellisario College of Communications. Her research interests include international news production of environmental issues, particularly in Spanish-language newsrooms. She is an award-winning documentary producer, and her research has appeared in *Science Communication*, *Communication Law and Policy*, *Journalism*, *Media History* and *Communication, Culture & Critique*, as well as in three books examining journalism and science communication in international contexts.

Paola Prado, PhD, is Associate Professor, Journalism and Digital Media, Roger Williams University. Her research on environmental risk news reporting and information and communications technology use for development and social change in Latin America appeared in various journals and in *Environmental News in South America: Conflict, Crisis and Contestation*. She holds a PhD in Communication from University of Miami, an MA in Latin American Studies from Georgetown University and a BA in Cinema from Denison University.

Anna Roosvall is Professor of Media and Communication Studies at Stockholm University. Her research is centred on justice, democracy and globalisation in linguistic and visual modes of communication. She researches cultural journalism and climate reporting. Her recent publications include *Media and Transnational Climate Justice* (2018, co-written with Matthew Tegelberg).

Chindu Sreedharan is Principal Academic in Journalism at Bournemouth University. A former journalist, he has a particular interest in journalistic storytelling as a means to improve human rights situations and empower marginalised groups. His research focuses on 'abnormal journalisms', reportage that extends the boundaries of conventional newswork—from crisis and post-disaster situations, on social media, as well as new forms of nonfictional narratives. He is co-principal investigator for Media Action Against Rape (MAAR). He tweets @chindu, blogs on https://chindu.co.uk/.

Matthew Tegelberg is an assistant professor in the Department of Social Science at York University. Situated at the intersection of media, culture and climate change, his research has appeared in *Journal of Science Communication*, *Nordicom Review*, *triple C: Communication, Capitalism & Critique* and several edited collections. He is also co-author, with Anna Roosvall, of *Media and Transnational Climate Justice: Indigenous Activism and Climate Politics* (2018).

Einar Thorsen, PhD, is Professor of Journalism and Communication and Executive Dean of the Faculty of Media and Communication at Bournemouth University. His research covers online journalism, citizens' voices and news reporting of crisis and political change—inextricably linked with protecting freedom of speech, human rights and civil liberties—especially for journalists, vulnerable people, marginalised groups and in contexts or countries where such liberties are being curtailed. He is joint principal investigator for Media Action Against Rape (MAAR). He tweets @einarthorsen.

Confidence Uwazuruike is an independent researcher and formerly of the Faculty of Media and Communication, Bournemouth University. His research focuses on the African media reporting of conflicts in Africa and on journalism and new media in Africa.

List of Figures

List of Tables

CHAPTER 1

Introduction: Media, Journalism and Disaster Communities

Jamie Matthews and Einar Thorsen

It is widely recognised that disasters are becoming more frequent, having greater and longer-lasting impacts and increasingly affecting those most vulnerable (Watson, Caravani, Mitchel, Kellet, & Peters, 2015). Climate-related and geophysical hazards—earthquakes, flooding and wildfires in particular—affect millions of people worldwide each year (Guhpa-Sapir, 2018). Disasters are also the consequence of human action or inaction, the failure of people to mitigate and respond to risks that arise from new technology, conflict and lack of governance, amongst others.

While those tasked with managing and responding to disasters, emphasise their event-driven nature and have sought to address hazards to reduce communities' exposure to disaster risks, there is now a greater recognition that disasters need to be defined and understood in broader terms, encompassing vulnerabilities that develop over time or persist due to structural conditions and inequalities for example. Poverty, environmental degradation, forced migration and conflict all contribute to insecurity (Beck, 2009). These insecurities can intersect with or aggravate the effects of

J. Matthews (✉) • E. Thorsen
Bournemouth University, Bournemouth, UK
e-mail: jmatthews@bournemouth.ac.uk; ethorsen@bournemouth.ac.uk

J. Matthews, E. Thorsen (eds.), *Media, Journalism and Disaster Communities*, https://doi.org/10.1007/978-3-030-33712-4_1

sudden-impact events. Disasters can therefore be more accurately conceptualised as complex, systemic failures (see Cottle, 2014).

With media and journalism integral to representations of and public communication about disasters, this book considers their role in the context of these new understandings of disaster. It adopts a unique approach by centring its analytical focus on what we term *disaster communities*, namely communities that are at risk from, affected by and recovering from the adverse impacts of disaster and their drivers and exploring their diverse relationships with media and journalism. The research, case studies and perspectives introduced in this collection consider how media and journalism produced by and for such disaster communities may offer alternative perspectives to national media, give voice to those vulnerable to hazards or seeking to rebuild after disaster, and also support risk reduction and recovery processes.

Disasters and Their Drivers

Disasters are very rarely solely related to sudden-onset events. Instead, their causes are complex, deep-seated and intersect with other vulnerabilities that create insecurity, most significantly environmental degradation and climate change, poverty, urbanisation and conflict (Sendai Framework for Disaster Risk Reduction, 2015). While there is a diversity of research across different disciplines, academic enquiry has tended to focus on disaster events and their acute impacts. Studies of media and journalism have generally followed this pattern (Ploughman, 1995), with greater scholarly attention paid to how media cover disaster and enable disaster communication (see Veil, 2012). The role of media in supporting risk reduction, enabling communities to identify and address the antecedent conditions that contribute to insecurity and intersect with disaster recovery processes are less well understood.

It is the emphasis on 'extreme events', often the calamitous (once in a generation) natural disasters that result in significant loss of life, that become known and made visible through international media coverage (Cottle, 2014), which has limited the scope of disaster research. Some scholars, therefore, identify a need for greater theoretical diversity in this body of research and for disaster studies to link to the related fields of sociology of risk and environmental sociology as well as to consider the key sociological concerns of inequality, diversity and social change (Tierney, 2007).

As indicated, a consensus has emerged in recent years that the frequency, impact and scale of disaster are increasing. The climate crisis is fuelling more powerful storms and prolonging periods of drought. Environmental degradation and the destruction of ecosystems, such as floodplains and forests, remove natural barriers that protect communities from hazards. Urbanisation is increasing the number of people exposed to natural hazards (Global Facility for Disaster Risk Reduction, 2016). Such hazards in the context of other conditions, threats and vulnerabilities that create or contribute to human insecurity, the most significant drivers being persistent poverty, food insecurity, forced migration, crime, conflict and violations or political and human rights, have changed and expanded understandings of disaster, their causes and impacts. Consequently, in policy and practice there has been a subtle shift away from disaster management towards disaster risk reduction, recognising the need to address underlying drivers of disaster. The Sendai Framework for Disaster Risk Reduction (2015, p. 10), implemented in 2015, illustrates this approach by calling for:

> more dedicated action focused on tackling underlying disaster risk drivers, such as the consequences of poverty and inequality, climate change and variability, unplanned and rapid urbanization, poor land management and compounding factors such as demographic change, weak institutional arrangements, non-risk-informed policies, lack of regulation and incentives for private disaster risk reduction investment, complex supply chains, limited availability of technology, unsustainable uses of natural resources, declining ecosystems, pandemics and epidemics.

These broader, multilayered approaches to understanding and defining disaster and their drivers have led to contemporary theorisations of disaster moving away from low probability yet high-impact processes to consider the accumulation and intersection of hazards, risks and vulnerabilities.

One approach that has been adopted in recent years is the concept of cascading disasters, which considers the interactions between compounding vulnerabilities and different events (Pescaroli, Nones, Galbusera, & Alexander, 2018). A cascading disaster may be initiated by a trigger event, either natural or anthropogenic, which then intersects with other hazards to exacerbate the impacts of this event and create or aggravate other vulnerabilities (Pescaroli & Alexander, 2015). With many vulnerabilities arising from increasing interdependence between systems, whether these are interactions between climatic, food and energy systems for example (Helbing, 2013), it is complexity and these interconnections that result in adverse outcomes.

The 2011 Japan disaster is often cited as an example of a cascading sequence of hazards and vulnerabilities. The primary trigger was an undersea earthquake off Japan's northeast coast, which generated a series of tsunami waves. In turn, the tsunami damaged cooling systems at the Fukushima nuclear plant and led to the release of radioactive material. It was the interactions between natural and technological hazards, therefore, that escalated the disaster and its impacts. Fears about widespread radioactive contamination, the extent of damage to Japan's infrastructure and its significance as a manufacturing base meant that its repercussions were not limited to the Asia Pacific region.

A cascading disaster is not simply a causative sequence of events and is more accurately described as a non-linear process that is context dependent. Therefore, a flood may result in loss of life, but the impact of flooding that occurs in a region or country where there is a greater reliance on subsistence farming and with weak housing and healthcare infrastructure, for example, would be magnified by the lack of economic and structural resilience. This amplification may create further adverse effects, such as triggering population movements or exacerbating intergroup tensions, which in turn may contribute to further secondary disasters, such as identity-based violence, conflict or other critical emergencies (Pescaroli & Alexander, 2015).

Other theorisations emphasise a broader paradigm, suggesting that a more appropriate focus for research are crises, as the 'exogenous and endogenous factors' that create disruption (Boin, 2005, p. 165). This perspective argues for a broader typology of disaster drivers to encourage interdisciplinary and multilevel approaches to analyse the causes of complex crises. It also recognises that adverse effects may return or create further unintended consequences even when a crisis has supposedly been resolved.

Quarantelli (2006, p. 9) describes newly emerging disasters that 'jump or cut across social systems' as trans-social-system ruptures (TSSRs). These ruptures are a consequence of globalisation processes and their effects are geographically dispersed. They are the major global crises, such as pandemics, transitional terrorism and climate change, that have widespread repercussions and require international and transnational solutions. These types of disasters are according to Cottle (2014, p. 4) 'endemic to, deeply enmeshed within and potentially encompassing in today's world disorder'. He argues that the contemporary media ecology and global communication flows have become increasingly important in how these global crises become signified, understood and responded to. Their visibility, how they are presented to audiences and global movements of public opinion are shaped by the dynamics of media and communication.

It is the recognition that disasters are a product of a range of complex hazards, risks and vulnerabilities that informs the selection of case studies and research that are brought together in this book. It includes examples that demonstrate the characteristics of traditional or older disasters as significant disruptive events (Quarantelli, 2006) but also introduces those that are shaped by development inequalities, arise from policy failures or intersect with other systemic risks, including climate change, environmental degradation, conflict and violence, amongst other hazards and vulnerabilities.

Despite these evolving understandings and their common characteristics, both *disasters* and *crises* are social constructions. What is defined or labelled as a disaster reflects values, interests and perspectives and encourages particular forms of intervention. Many events and drivers of disaster that result in adverse impacts, with significant loss of life and widespread destruction, fail to register for international media. News values, the proximity and relevance or emergencies to audiences and their potential to dislocate the interests of elite nations render some 'disasters' and emerging crises invisible (Galtung & Ruge, 1965; Joye, 2010). Social construction processes are also shaped by media treatment of events and how risks and vulnerabilities are reported. Tierney (2007, p. 62) notes that in covering disasters media 'both reflect and reinforce broader societal and cultural trends, socially constructed metanarratives, and hegemonic discourse practices that support the status quo and the interests of elites'. This, she argues, was evident in the way the US media constructed post-Katrina New Orleans as lawless and violent, which drew upon stereotypical portrayals of America's impoverished communities and also reflected longstanding political discourses and policy positions that promote a greater role for the military in disaster management.

Alternatively, it can also be argued that adopting the term 'disaster' to describe slow-moving and chronic vulnerabilities, drivers for disaster that may not have yet reached the acute stage, serves to call attention to a problem that may not be recognised or understood as such. The rhetorical dimension of identifying a hazard, risk or vulnerability, or indeed their potential to emerge, and their intersection with others as a 'disaster' may encourage responses that attempt to address or alleviate these drivers of adverse impacts for communities. There is, however, a danger that *disaster* becomes a catch-all term and by encompassing a range of disruptions, hazards and risks will lead to a loss of analytical precision. Yet, as it has been argued, there is a recognition in contemporary theorisations that *disaster* should be redefined to reflect the accumulation and intersection of hazards, risks and vulnerabilities that lead to adverse effects.

Media, Journalism and Disaster Communities acknowledges these broader understandings of disaster. Alongside the well-established natural and anthropogenic hazards that create disaster vulnerabilities, this book also presents research and case studies that call attention to other drivers for disaster. Importantly, as is elaborated on further below, it seeks to emphasise community perspectives towards disaster. Ultimately, the definition and understandings of disaster and their effects, ontologically, is located in the different notions of community and how these communities may view and understand disaster and their effects.

Defining Disaster Communities

Community is a broad concept that is used to describe groups of people with social connections and/or shared commonality, for example through some form of shared identity, religion and customs or values, typically in a defined geographical locality and a shared sense of belonging. Geographic specificity contributes to the boundary definition of communities, through shared places such as towns and neighbourhoods, though community can also be used in connection with the expression of national or international identities. As such, communities vary significantly in scale—both in terms of their population and in terms of their physical distribution. Globalisation has contributed to a shift in how we view communities, away from the need for physical colocation in order to form community ties to the formation of diaspora communities and online communities. Physicality is also not necessarily fixed, with nomadic communities exercising mobility as part of their core identity, and refugees or migrants often retaining a sense of belonging both to their place of origin and to each other despite displacement.

Indeed, for the purpose of this book, we embrace broad and multifaceted understandings of communities—from towns and villages to regions or states within a nation, from fixed geographical location to displaced peoples, from physical colocation to online connectivity, from kinship to professional or religious ties—and recognise that individuals may belong to a number of such communities at different levels at any one point.

The purpose of community is often seen as a way of uniting groups of people, and that such unity reduces suffering through solidarity, collective endeavour and shared purpose. Members of communities strive for well-being and shared emotional wellbeing (Davidson & Cotler, 1989; MacMillan & Chavis, 1986). Disasters disrupt the fabric of communities, regardless of their size, and necessarily trigger a response from its members to protect, address vulnerabilities and restore normalcy of its existence. Irrespective of their causes, disasters, Bruhn (2011, p. 112) argues, "have common effects—they produce trauma that changes the social and emotional lives of the individuals, the resiliency of families, and the cultural fabric of families." In so doing, they can also give rise to new communities that are defined by people, for example survivors of or displaced by a disaster, crisis or conflict.

Defining disaster communities solely based on those residing in a geographic area physically damaged or disrupted by disaster is also limiting because those affected may expand beyond this, which we refer to as secondary communities of concern. That is, the impact on family or friends living in other locations, people living close to affected areas who experience increased anxiety and fear as a consequence of *perceived* risk or those that have fostered connections to those communities directly affected by disaster. Here we concur with Kirschenbaum's (2004, p. 98) assertion that a disaster community has "a specific geographic disaster epicentre but is perceived and experienced through a complex web of social networks". Kirschenbaum argues "the foundation of a disaster community depends on a core of social networks connecting those directly or indirectly affected by a disaster" (Ibid.). This conceptualisation of disaster communities acknowledges that there are a number of "social networks operating simultaneously from the epicentre of a potential (or actual) disaster area" (Kirschenbaum, 2004, p. 100). Components of a disaster community according to Kirschenbaum are defined as family network, micro-neighbourhood network, and macro-community network. Whilst these provide focal points for understanding levels of preparedness, they do not consider professional networks beyond emergency services. That is, the role of *local* or *community* media in developing community identity to raise awareness of risks and vulnerabilities, their role in providing disaster warnings for example, and their contribution to rebuilding community ties and social relations post-disaster. In this book, we argue not just for a more expansive definition of disaster communities to include such communicative dynamics, but contend that the dynamics and composition of

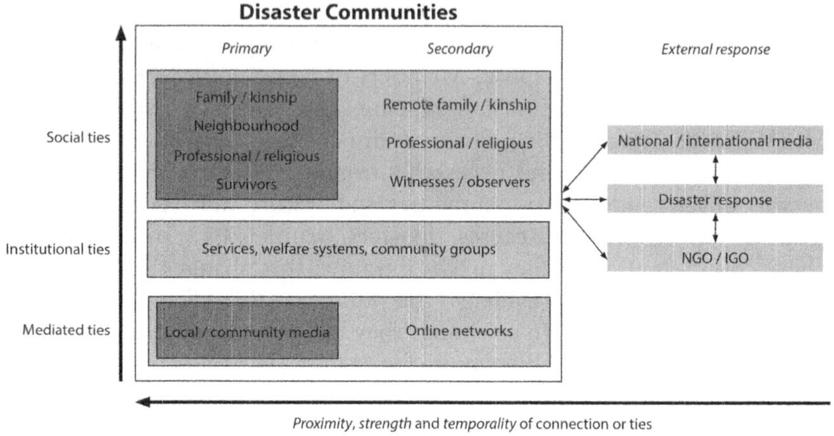

Fig. 1.1 Model of disaster communities

disaster communities cannot be understood in isolation from the epistemology of mediated interrelations.

When we define ***disaster communities*** as a concept, we draw on how the response to disaster helps solidify the sense of community and its boundaries—be that as a community at risk from, impacted by, or recovering from disaster. Moreover, spontaneous connections arise as a response to disasters, for example from emergency service response, non-governmental organisation (NGO) and volunteer networks and through coverage by national and international news media. These create new points of connection for both social and mediated ties within disaster communities. Social and mediated ties, as understood in our conceptualisation of disaster communities, have a greater degree of permanency and provide both boundary definitions and identity for these communities. This is partly because such ties have pre-existing connections, and because the communities have a shared experience or purpose through their exposure to vulnerabilities, as survivors or in recovery. The different connections and their proximity, strength and temporality for disaster communities are illustrated by Fig. 1.1.

Due to the way we conceptualise disaster communities, this book has a particular emphasis on local media and journalism, which as part of the social and civic fabric of their communities are able to raise awareness of disaster drivers to provide information to support communities to mitigate risk and prepare for disaster (Blanchard-Boehm, 1998). When communication infrastructures are disrupted, such as may follow a significant 'natural disaster', often it is only local media—for example community radio—that are able to meet communities' information needs, providing access to vital emergency information (Kanayama, 2007). Later, beyond the immediate impacts, local media and journalism are able to support communities as they seek to adapt to and recover from disaster, for instance by advocating on their behalf and raising awareness of the issues of post-disaster recovery (Matthews, 2017; Usher, 2009). This is reflected in the close working relationships humanitarian agencies seek to develop with local media, where there is a recognition of the importance of capacity-building to support local media in recovering communities.

Media, Journalism and Disaster Communities also recognises how, in the context of disasters and disaster communication, the media environment is evolving, with digital media facilitating greater public participation in journalism and information provision, offering new opportunities to serve audiences, those that may coalesce around particular interests, and enhancing the breadth of media and journalism accessible to these audiences. The aggregation of social media data is now integral to disaster planning, used to warn people about hazards and support emergency response (Crowe, 2012). Following disasters, people may use online platforms to create their own blogs and websites bypassing traditional gatekeepers to communicate the realities of the post-disaster situation in their communities (Farinosi & Treré, 2014). Other forms of 'hyperlocal journalism', by providing information or news for a small, defined community, are able to raise awareness of hazards that have not yet come to the attention of mainstream media. New media are also changing the dynamics of humanitarian communications, allowing direct connections between affected communities and humanitarian agencies and facilitating the emergence of networks of digital volunteers that are able to support emergency response and assist in rescue and relief efforts (Chernobrov, 2018).

By defining *disaster communities* in a new and more expansive way, and also including local and community media as important ties, it provides greater analytical precision to the communicative dynamics and social interconnections that emerge from hazards, vulnerabilities and risks that drive disaster. This definition also draws attention to an often-neglected area of research—the role of local and community media in relation to disasters—which serves as a bridge to media research that has largely considered the processes of mediation at a national and international level, sociological research that has focused more on social ties than those attributed to mediation, and disaster management research that has emphasised risk reduction, preparedness of national agencies and information provision in emergencies.

Overview of Chapters

To expand and elucidate the concept of disaster community and its intersections with media and journalism, the book presents 15 chapters. They coalesce around the different characteristics and points of connection of disaster communities outlined above and together address the following key questions guiding this edited collection: What are the different forms of media and journalism produced by and for communities at risk from, affected by and recovering from disaster? To what extent do they recognise and speak to the different notions of community that emerge in disaster contexts? How may news and media content be received and acted upon by these communities? Moreover, to consider to what extent media and journalism may help enhance our understanding of the breadth of hazards, risks and drivers of vulnerability? Within the context of the complex and globalised nature of such disaster, what contribution can community approaches make to addressing these vulnerabilities?

The book is divided into three parts that reflect different types of hazards and drivers of disaster that may have adverse impacts on communities. Part I, *Environmental Destruction and Geophysical Hazards*, introduces five chapters that emerge from the vulnerabilities and consequences for disaster communities arising from environmental destruction and geophysical hazards. Our opening chapter by Paola Prado and Juliet Pinto, evaluates how local and national media covered the collapse of the tailings dam at the Samarco mine and the subsequent contamination of the Doce River basin in Minas Gerais, Brazil. This mining disaster further underlined the risks for communities in proximity to these metal and ore mines and their wider environmental impacts. By comparing reporting in the *Estado de Minas*, a newspaper published in Minas Gerais state, and *Folha de São Paulo*, a national daily title, they consider how local media defined the disaster, articulated community agency, vulnerability and resiliency. They argue that as demand increases for energy and mineral resources in Latin America, it's vital that communities receive news that goes beyond official narratives and offers in-depth community-orientated discussions that are able to foster resilience. The significance of this disaster risk for communities was sadly brought into focus again in January 2019 following the collapse of another Vale tailings dam at Córrego do Feijão in Minas Gerais that left more than 200 people dead.

Chindu Sreedharan and Einar Thorsen centre their analysis on journalists as a professional community and consider how reporters negotiate their professional and personal responsibilities when they are affected by disaster. By analysing the experiences of journalists in Nepal after the earthquake of April 2015, they argue that journalists' dual status created a clash between their personal and professional selves that had to be constantly negotiated in part as a consequence of obligations to their overlapping *afno manche* kinships/networks. Moreover, these difficulties also influenced their reportage, and engendered a strong interventionist commitment to their communities and what they could do for those in need, and also their perceptions of disaster journalism and expectations of what is required to help rebuild post-disaster societies.

In a similar vein, Jamie Matthews explains how after significant disasters secondary, interest-based communities may emerge. This is illustrated through the evaluation of two grassroot media initiatives that were launched after the 2011 Japan tsunami that sought to share news and information in English about recovering communities in Miyagi prefecture. These two projects, he argues, while also seeking to support recovery and reconstruction highlight the ongoing post-disaster issues in Tōhoku, and more specifically those experienced by people living in the cities of Ishinomaki and Kesennuma, are also able to meet the information needs of a broader community of concern, one that is geographically dispersed but shares a connection to those directly affected by disaster.

Jacqui Ewart's chapter documents research that has assessed the role of the news media in Australia when covering disasters. By considering the different natural hazards and their intersection that drive disasters in Australia, and specifically by discussing examples of recent droughts, Ewart identifies the valuable information that local and community news media are able to provide to help communities deal with disaster. This is despite the fact that organisations engaged in disaster management are increasingly turning to social media to distribute information about these disaster risks.

The challenges posed by climate-induced migration, a systemic risk that cuts across territories, for island communities is the focus for Anna Roosvall, Matthew Tegelberg and Florencia Enghel's chapter. Their analysis introduces three US islands, Sarichef Island in Alaska, Isle de Jean Charles in Louisiana, and Puerto Rico, as cases that, while of contrasting geographical size, share similarities and differences in their degrees of poverty, the rights of indigenous populations and political representation. They explore how imminent climate migration from these islands is understood in local and transnational reporting, arguing that different issues are brought to the fore in both. Transnational journalism, illustrated by articles carried by the *Guardian*, tends to reflect global perspectives, which are significant to recognise how climate change serves as a driver for migration. Local journalism, however, places greater emphasis on the needs of affected communities and connections with other issues, such as infrastructure and political processes. Both, they conclude, are necessary to gain a complete picture of climate migration and its impacts on vulnerable communities.

Part II brings into focus four case studies that coalesce around the theme of *Armed Conflict and Journalistic Freedoms.* Significantly, how persistent low-intensity conflict, the challenges of post-conflict recovery and restrictions on journalistic freedoms contribute to insecurity for disaster communities. It begins with a chapter by Mathew Charles which evaluates how El Faro, an online news portal, has been able to challenge the dominant narrative of urban violence in El-Salvador by incorporating the voices of gang members into their coverage. His analysis shows that including the perspectives provided by perpetrators of violence enables a deeper understanding of the violence and its underlying causes, which can help to repair the social bonds in communities that have been broken by gang violence.

The conflict and its impacts on communities in the Donbas region of Ukraine is introduced in Dariya Orlova's chapter. She presents a study of Donbas residents' news consumption, identifying their frustrations with and distrust in mainstream media coverage of the Ukraine-Russia conflict. She argues that conflict-affected communities did not consider local media as a reliable source for information and that their coverage contributed to the frustration, resentment and disconnectedness they felt. Instead, local and community groups on social networks were found to offer an alternative to local news media and were considered more effective in meeting communities' information needs.

The characteristics of local news coverage of the farmer-herder conflict in Nigeria are explored in Confidence Uwazurike's chapter. Through a longitudinal analysis of coverage in *Idoma Voice*, a local newspaper that caters for a community disrupted by the conflict, he identifies how reporting has tended to emphasise the violent aspects of the conflict due to its impact on and relevance to its readership. The consequence of rendering such frames is that they fail to promote alternative strategies and encourage non-violent solutions to the conflict. This, he suggests, offers a case for developing a model of peace journalism that is particular to local and community media.

Aida Al-Kaisy examines developments in Mosul following its liberation from *Islamic State* in 2017 and the role of different local media, specifically radio and online, in post-conflict reconciliation. She writes that post-conflict recovery for Mosul is limited by the political, social and economic conditions that persist and this is reflected in how media operate within this landscape.

Our third and final part, *Human (In)action and Humanitarian Crises*, presents four chapters that introduce examples of disaster and their drivers that result from human (in)action, including governance and policy failures, or are a consequence of or have evolved to become humanitarian crises, specifically communities that are displaced or fleeing persecution. Kurt Barling focuses on the Grenfell Tower fire that occurred in Kensington, London, in 2017. He explores how local journalism failed to report on Grenfell before the blaze, ignoring the publicised concerns of the community, in particular those raised by the Grenfell resident's association about fire safety in tower blocks that followed a similar fire at Lakanal House in Camberwell, South London, in 2009. In this sense, he argues, Grenfell was a disaster foretold.

Kyle J. Holody compares national and community newspapers coverage of the mass shooting that occurred at Marjory Stoneman Douglas High in Parkland, Florida, in 2018. He demonstrates that community newspapers were far more likely to cover the individual, community and regional ramifications of the shootings than national newspapers, which focused much more on the societal/international aspects of the shootings. This, Holody, argues, combined with the greater amount of coverage they afforded to the event and its impacts, suggests that community newspapers are able to keep attention on this issue after national media have moved on to other stories.

The refugee community in Greece provides the focus for Victoria Jack's chapter. She evaluates a humanitarian information project, News that Moves, which sought to enhance refugees' access to reliable information that would help them navigate legal procedures and access assistance.

Jack's analysis identifies how the News that Moves website and then later Facebook page became an important and trusted source for refugees as they travelled from Greece and through the Balkan route into Western Europe. She also acknowledges how the political sensitives and security concerns may impinge on the ability of such initiatives to meet communities' information needs, which ultimately may reduce refugees' perception of them as a trusted information source.

Lisa Brooten examines the role of social media, Facebook in particular, in the violence perpetrated towards the Rohingya community in Myanmar. Through interviews conducted with members of civil society organisations in Myanmar, she considers how these organisations were able to raise concerns about Facebook and the spread of hate speech and incitement of violence towards the Rohingya. Brooten argues that in authoritarian contexts such as Myanmar, it is vital for companies such as Facebook to work closely with local partners to identify problematic actors that seek to create community division and to support communities to expand their options for communication, including developing alternative platforms to Facebook.

The book concludes with an afterword by Mervi Pantti that draws together the central themes of the book to elucidate our understanding of *disaster communities* and the integral role of local and community media within these.

We hope that the case studies, research and perspectives presented in this book provide an eclectic overview of how different forms of media and journalism contribute to our understanding of the lived experiences of communities at risk from, affected by and recovering from disaster. In so doing, these contributions illustrate the utility of *disaster communities* as an analytical concept for future research and, crucially, the importance of stepping beyond national or international news media to consider media and journalism produced by and for disaster communities.

References

Beck, U. (2009). Critical theory of world risk society: A cosmopolitan vision. *Constellations, 16*(3),22.https://doi.org/10.1111/j.1467-8675.2009.00534.x

Blanchard-Boehm, D. R. (1998). Understanding public response to increased risk from natural hazards: Application of the hazards risk communication framework. *International Journal of Mass Emergencies and Disasters, 16*, 247–278.

Boin, A. (2005). From crisis to disaster: Towards an integrative perspective. In R. W. Perry & E. L. Quarantelli (Eds.), *What is a disaster* (pp. 153–172). International Research Committee on Disasters.

Bruhn, J. G. (2011). *The sociology of community connections*. Springer Science & Business Media.

Chernobrov, D. (2018). Digital volunteer networks and humanitarian crisis reporting. *Digital Journalism, 6*(7), 928–944. https://doi.org/10.1080/21670811.2018.1462666

Cottle, S. (2014). Rethinking media and disasters in a global age: What's changed and why it matters. *Media, War & Conflict, 7*(1), 3–22. https://doi.org/10.1177/1750635213513229

Crowe, A. (2012). *Disasters 2.0: The application of social media systems for modern emergency management*. CRC Press.

Davidson, W. B., & Cotler, P. R. (1989). Sense of community and political participation. *Journal of Community Psychology, 17*, 119–125.

Farinosi, M., & Treré, E. (2014). Challenging mainstream media, documenting real life and sharing with the community: An analysis of the motivations for producing citizen journalism in a post-disaster city. *Global Media and Communication, 10*(1), 73–92. https://doi.org/10.1177/1742766513513192

Galtung, J., & Ruge, M. H. (1965). The structure of foreign news. *Journal of Peace Research, 2*(1), 64–90.

Global Facility for Disaster Risk Reduction. (2016). *The making of a riskier future: How our decisions are shaping future disaster risk*. Retrieved July 1, 2019, from https://www.gfdrr.org/sites/default/files/publication/Riskier%20Future.pdf.

Guhpa-Sapir, D., (2018). 2018 review of disaster events. Centre for Research on the Epidemiology of Disaster. Retrieved July 1, 2019, from https://cred.be/sites/default/files/Review2018.pdf.

Helbing, D. (2013). Globally networked risks and how to respond. *Nature, 497*, 51–59.

Joye, S. (2010). News discourses on distant suffering: A critical discourse analysis of the 2003 SARS outbreak. *Discourse & Society, 21*(5), 586–601. https://doi.org/10.1177/0957926510373988

Kanayama, T. (2007). Community ties and revitalization: The role of community radio in Japan. *Keio Communication Review, 29*(3), 5–24.

Kirschenbaum, A. (2004). Generic sources of disaster communities: A social network approach. *International Journal of Sociology and Social Policy, 24*(10/11), 94–129. https://doi.org/10.1108/01443330410791073

MacMillan, D. W., & Chavis, D. M. (1986). Sense of community: A definition and theory. *Journal of Community Psychology, 14*(1), 6–23.

Matthews, J. (2017). The role of a local newspaper after disaster: An intrinsic case study of Ishinomaki, Japan. *Asian Journal of Communication, 27*(5), 464–479. https://doi.org/10.1080/01292986.2017.1280065

Pescaroli, G., & Alexander, D. (2015). A definition of cascading disasters and cascading effects: Going beyond the "toppling dominos metaphor". *Planet @ Risk, Global Forum Davos, 3*(1), 58–67.

Pescaroli, G., Nones, M., Galbusera, L., & Alexander, D. (2018). Understanding and mitigating cascading crises in the global interconnected system. *International Journal of Disaster Risk Reduction, 30*(B), 159–163.

Ploughman, P. (1995). The American print news media 'construction' of five natural disasters. *Disasters, 19*(4), 308–326.

Quarantelli, E. L. (2006). The disasters of the 21st century: A mixture of new, old, and mixed types. *Disaster Research Centre Preliminary Papers*, 353. Retrieved July 1, 2019, from http://udspace.udel.edu/handle/19716/2374.

Sendai Framework for Disaster Risk Reduction. (2015). Retrieved July 1, 2019, from https://www.unisdr.org/files/43291_sendaiframeworkfordrren.pdf.

Tierney, K. J. (2007). From the margins to the mainstream? Disaster research at the crossroads. *Annual Review of Sociology, 33*, 503–525.

Usher, N. (2009). Recovery from disaster: How journalists at the New Orleans Times-Picayune understand the role of a post-Katrina newspaper. *Journalism Practice, 3*(2), 216–232. https://doi.org/10.1080/17512780802681322

Veil, S. R. (2012). Clearing the air: Journalists and emergency managers discuss disaster response. *Journal of Applied Communication Research, 40*(3), 289–306. https://doi.org/10.1080/00909882.2012.679672

Watson, C., Caravani, A., Mitchel, T., Kellet, J., & Peters, K. (2015). Financing for reducing disaster risk. UN Development Programme. Retrieved July 1, 2019, from https://www.odi.org/sites/odi.org.uk/files/odi-assets/publications-opinion-files/9480.pdf.

PART I

Environmental Destruction and Geophysical Hazards

CHAPTER 2

Brazilian Local and National News Coverage of the Samarco Disaster: A Disaster for the Community, the Corporation or the Environment?

Paola Prado and Juliet Pinto

On 5 November 2015, the Samarco dam holding by-products of iron mining waste collapsed, leaving 19 people dead and hundreds homeless. It is widely considered to be one of Brazil's worst environmental disasters, as a toxic mix of mud, sediment, and contaminants swept through villages, into the Doce River basin, and eventually out into the Atlantic Ocean. Beyond the immediate loss of life and destruction of property, the collapse created longer-term health, environmental, and economic difficulties for the communities who depend on the Doce River and its tributaries for drinking water, crop irrigation, fishing, and manufacturing. The contamination spread through river and tributary ecosystems and water supplies

P. Prado (✉)
Roger Williams University, Bristol, RI, USA
e-mail: pprado@rwu.edu

J. Pinto
Pennsylvania State University, University Park, PA, USA

J. Matthews, E. Thorsen (eds.), *Media, Journalism and Disaster Communities*, https://doi.org/10.1007/978-3-030-33712-4_2

were suspended in cities, towns, and villages throughout the state of Minas Gerais and beyond as tons of mining debris flowed down river on its trajectory to the estuary in the coastal state of Espirito Santo. Prior to the dam collapse at Córrego do Feijão in Minas Gerais in January 2019, the Samarco breach was the largest environmental mining disaster caused by tailings dam failure (Carmo et al., 2017; Hatje et al., 2017). It halted operations for Samarco, a Brazilian mining giant that is a joint venture between the Brazilian Vale group and Anglo-Australian multinational BHP Billiton, leaving many unemployed. The company, which in the 1970s scaled up mining operations, is also a major political donor (the fifth largest in Brazil) and a primary source of employment (Ruhfus, 2017) in Minas Gerais, a name that translates to English as "General Mines."

At the time of writing, now four years on from the disaster, the political fallout and concerns over the cleanup—already estimated to be a decade-long undertaking—continue. Affected communities have questioned the cleanup strategies, complaining that Samarco have planted vegetation over contaminated soils and have failed to remove toxic mud. Furthermore, Brazilian environmental authorities published a report noting that Samarco had rerouted various rivers and tributaries (Phillips, 2016). In 2018, the mining company announced that limited operations would restart in 2020 and that a settlement had been reached with Minas Gerais prosecutors (Mining Technology, 2018).

As in many such events, media accounts of the disaster informed international audiences of "what happened." How regional media translate the dispute, position the actors, and provide the perspectives is important for national and international media coverage, and vice versa (Funk & McCombs, 2017). At a time when left-wing governments in Latin America turn to China for new sources of capital, a commodity boom powered by new mega-dams has challenged newsrooms in the region to accurately report upon the steady proliferation of large extractive projects in faraway sacrifice zones (Casey & Krauss, 2018; Pinto, Prado, & Tirado-Alcaraz, 2017).

In this case study of local and national news coverage, we combine corpus linguistic analysis with qualitative news frame analysis to examine the lexical relationships and thematic framing of environmental emergencies, risk mitigation, recovery, and adaptation in the regional paper *Estado de Minas* (Minas Gerais) and the national paper *Folha de São Paulo*. We explore the themes, trends, clues, and strategies observed in media coverage and compare them in terms of regional or national coverage, all within the context of understanding mediated constitutive roles. We ask how these findings are significant for the flows of political action and rhetoric and, ultimately, the transnational political and public spheres.

Local News and Disaster Coverage

Local news coverage provides an important lens through which to observe coverage of environmental disasters and crises, as the immediate impacts are often experienced most severely by residents at the local level, and local news can be the main source of information for these affected communities. The idea of "local" media can be complicated, as Funk and McCombs (2017, p. 846) write, "the word 'local' has a complicated identity…. Local media play a major role in articulating the inclusions and exclusions that define culture, language and place." For local and regional news outlets, disasters that impact on their audiences can drive extensive coverage, with particular sourcing patterns and angles that differ from national coverage (Branton & Dunaway, 2009; Hamilton, 2004; Molotch & Lester, 1975; Ploughman, 1995).

Such news values can also provide openings that otherwise may not make it into coverage. For example, invisible, incremental, and complex topics such as climate change may receive more coverage from local media, as audiences experience them firsthand. In the example of reporting on rising sea levels in regional media in South Florida, coverage can be influenced by flooding events, local and state political debate, and community events (Jacobson, Pinto, Gutsche, & Wilson, 2019, in press), as well as by newsroom entrepreneurs who push stories out past organizational gatekeepers (Pinto & Vigon, 2018). Conversely, local news coverage of sea level rises can mimic national news coverage by focusing on the rhetoric of local, state, and national politicians, rather than the actions and voices of local community members (Gutsche, Jacobson, Pinto, & Michel, 2017).

Local news coverage of environmental issues can be complicated by structural and institutional variables. As circulations and readerships decline across newsrooms, constraints on budgets and personnel can mean that national organizations are not always able to send correspondents to cover issues "on the ground"; this can be particularly important for events in rural areas with a weak institutional rule of law (Pinto et al., 2017). Practicing journalism in many parts of the world can carry significant risks, particularly in Latin America: Freedom House cited Brazil among the world's most dangerous places for journalists (Abramowitz, 2017). The practice of environmental journalism in Latin America can present particular challenges, including a lack of financial support, training, specialized knowledge, or protection from threats, as well as punitive legal environments for those investigating corporate or official corruption (Hughes & Lawson, 2005; Pinto et al., 2017; Waisbord, 2000).

However, when disasters strike, local news coverage can provide insights into political and social machinations. How local media communicate events, develop the narrative, define risk and translate political response has significant implications for public understanding and sometimes policy outcomes. Research has examined patterns of framing, authority response, sourcing, dramatization, and the cyclical nature of coverage (e.g. Houston, Pfefferbaum, & Rosenholtz, 2012). In the case of Hurricane Andrew, Salwen (1995) found that local and national papers quoted state officials and citizens, but national papers quoted national actors much more frequently. Mourão and Sturm (2018) analyzed Brazilian media coverage of the Samarco disaster and found that coverage was episodic, relied on official sources, provided little context, or seldom mentioned environmental damage.

Yet, less is known about local news coverage of environmental disasters beyond the immediate coverage (Gant & Dimmick, 2000). Impacts beyond the immediate aftermath of an event can linger, and political response can be drawn out and convoluted. In particular, how local news coverage may articulate concerns about (re)building communities after disaster, address questions of resiliency and risk, and define the disaster itself? While episodic coverage moves on from the disaster, their impacts on affected communities remain. In the context of politicization, risk, and damage to health and environmental well-being, how do local and national news coverage of a disaster compare?

The Brazilian Context

The world's top producer of iron ore, pellets, and nickel, and Brazil's largest private company, Vale, operates in over 30 countries; valued at US$82 billion, it posted annual revenues of US$34 billion in 2018. Its vast mineral reserves in the Brazilian Amazon include the open-pit Carajás complex, the world's largest iron ore mine, and the largest gold reserve in Brazil, the bountiful Serra Pelada.

The possibility of collapse of the Fundão tailings dam were not unknown. Experts had previously warned company officials of the possibility of ruptures to the dam's safety walls due to modifications (Freitas,

Silva, & Menezes, 2016) that compounded original construction flaws (Carmo et al., 2017). Residents also reported fears over a potential collapse, pollution, and other perceived threats before the collapse (Zhouri et al., 2016).

When the dam failed in the afternoon of 5 November 2015, an estimated 50 million tons of toxic mine waste flooded a nearby town and traveled more than 650 kilometers to the Atlantic coast via the Doce River. The spill devastated the watershed's ecosystems, threatened water supplies, and severely impacted local residents, particularly farmers and fishermen (cited in Agurto-Detzel et al., 2016).

The disaster was considered to be one of the most significant in terms of damage to human and environmental health (Zhouri et al., 2016) and made international headlines. Yet, in Brazil, national media coverage of environmental crises outside major cities has been uneven, with the exception being some instances of multimedia, in-depth coverage from *Folha de São Paulo* (Pinto et al., 2017). This chapter, therefore, provides a deeper exploration of the nuances between local and national media coverage of the Samarco disaster by examining reporting in two news dailies: *Folha de São Paulo* and *Estado de Minas*. We ask how local media defined the disaster, presented community agency, victims, vulnerability, and resiliency, and how this compared to national coverage over time.

Methods

Two privately held major news dailies, one national, *Folha de São Paulo*, and one local, *Estado de Minas*, present a basis for comparison of news coverage of the dam collapse in Mariana, where local communities, multinational conglomerates, regional and national political players, and other stakeholders struggled in the aftermath of the environmental crisis caused by the dam rupture. Historically, both newspapers have aligned themselves with the ruling government, yet both turned against the democratically elected administration of President Goulart in 1964, to support a military coup orchestrated in Minas Gerais, backed by the United States, and endorsed by Brazilian mainstream media. Both dailies backed the restoration of democracy in 1984, yet their support waned throughout the left-leaning administrations of President Luiz Inácio Lula da Silva (2003–2011) and his successor and former chief of staff and minister of mining, Dilma Rousseff, who was impeached and removed from office in 2016.

Established in 1921 and controlled by the Frias family since 1962, the privately held *Folha de São Paulo* is the national daily with the largest print circulation in Brazil, estimated at 351,917 (*Meio & Mensagem*, 2017). *Estado de Minas* is the property of privately held Brazilian broadcast, print and digital media conglomerate Grupo Diários Associados. Established in 1928 and headquartered in the state capital, *Estado* remains the most influential newspaper in Minas Gerais and one of Brazil's largest news dailies with an estimated weekday circulation of 63,613 and 85,496 on Sundays (*Meio & Mensagem*, 2017).

Our content analysis focused on news articles published online. In the manner of Quandt (2008), this sampling method reflects an equivalence for print and replicates Hsieh and Shannon's (2005) method of counting keywords previously identified in the literature, followed by textual interpretation that amplifies quantitative analysis. An online search for keywords "Samarco," "barragem de Fundão," "BHP Billiton," and "Fundação Renova" returned 2085 items (760 in *Folha* and 1325 in *Estado*). From this sample we produced a random constructed subset sample of 100 articles—29 articles from *Folha de São Paulo* and 71 from *Estado de Minas.* The analyzed articles range from 5 November 2015, the date the dam collapsed, through to 31 December 2017, filtered to retain only news articles published in the main section and topical news sections on environment, social entrepreneurship, and, in *Estado de Minas*, to include a special section about the dam collapse, "Vozes de Mariana" (Voices of Mariana).

On the constructed subset sample, AntConc, a corpus linguistic tool, was used to identify patterns among groups of words, analyze word positions, probabilities, placements, and associations, which was complemented by NVivo analysis of word trees and word frequencies. This was followed by a qualitative news frame analysis of the first three paragraphs of each article to identify evidence of emerging frames, the primary focus of coverage and use of sources.

Results

In the aftermath of the dam collapse, *Estado de Minas* referred consistently to Samarco and Vale yet seldom mentioned the foreign conglomerate BHP Billiton whose name, which appears in headlines, failed to return results in the *Estado de Minas* search engine. The analysis shows frequent mentions of Samarco in the two months after the dam collapse ($N = 94$)

and for two years after, more so as legal actions mounted (N = 220 in 2016 and N = 172 in 2017). Fewer mentions of the dam itself and foreign corporations indicate the disaster was closely identified with local company Samarco, rather than its multinational partners.

Folha's news coverage relied primarily on staff deployed to the scene, rather than wire agencies (17 percent). In contrast, three months after the collapse, as suspended particles still tainted the river red and metal ore soiled the sediment (Hatje et al., 2017), *Estado* shifted to wire reports for 43 percent of its coverage. Both publications relied primarily on episodic reporting (64 percent in *Folha* and 70 percent in *Estado*). Overall, longer, more in-depth reporting that provided information beyond the immediate crisis and official narratives of response to the event was limited to 24 percent of news articles. Only six stories were investigative in nature.

NVivo analysis shows frequent mention of "justice," yet scarce reference to those directly impacted by the collapse. The words which most frequently associated the event to environmental disaster emerged in the first 26 months and include: "disaster" (N = 88), "mineral" (N = 53), "mining" (N = 39), "risk" (N = 41), "recovery" (N = 41), "victims" (N = 29), and "indemnity" (N = 18) (see Table 2.1). In that time period, mentions of "mining," "minerals," "disaster," and "deforestation" doubled, the latter not mentioned in the immediate aftermath. Throughout those two years, use of the word "recovery" tripled in use (N = 18 and N = 17), while "risk" (N = 13, N = 15) remained constant and "victim" subsided (from N = 18 in 2015 to N = 10 in 2017). The word "destruction" spiked in the news coverage in 2016 (N = 10) while "emergency" subsided from eight mentions in 2015 to none in 2017. Absent or rare are mentions of the root words or derivatives for "contamination," "pollution," "consequences," "adaptation," "mitigation," "sustainability," or "vulnerability."

News Frame Analysis

Overall, the news coverage focused on the economic impact in the state: the *Estado de Minas* characterized the 15-meter-high toxic wave that engulfed 153 homes and structures in the village of Bento Rodrigues as a "mud tsunami" which left a "stain on the economy" of the rural and mining area where small ranchers and industry rely on the water of the Doce River. Event-centered reporting followed a narrative arc from initial victim testimonials, descriptions of the disaster and risk assessment to

Table 2.1 Frequencies of words associated with the Fundão Dam collapse

	2015	*2016*	*2017*	*Total*
Disaster	20	20	48	88
Minerals	14	13	26	53
Recovery	6	18	17	41
Risk	13	13	15	41
Mining	12	3	24	39
Victims	18	11	0	29
Destruction	4	10	7	21
Indemnity	5	3	10	18
Deforestation	0	6	10	16
Pollution	3	7	5	15
Crisis	2	9	3	14
Contamination	3	4	5	12
Devastation	3	3	5	11
Emergency	8	3	0	11
Homeless	6	0	4	10
Sustainability	3	1	5	9
Urgency	4	1	3	8
Conservation	1	1	2	4
Contingency	2	0	0	2
Regeneration	2	0	0	2
Consequences	0	0	1	1
Mitigation	0	0	1	1
Vulnerability	0	0	1	1
Adaptation	0	0	0	0

narratives about recovery efforts, and, later, various issues related to indemnification and delays in reparation. However, none provided long-term critical analysis of fixing lax and ineffective environmental regulations. Within 10 days of the dam break, congressional representatives revisited a stalled mining code and reassessed requirements that mining companies secure risk insurance to cover environmental damages. And 20 days after the disaster, the state assembly approved a controversial law that fast-tracked environmental permits for mining operations and removed oversight from a commission that included representatives from civil society, moving decision-making to appointed government officials.

In the immediate aftermath of the dam break, *Estado* employed frames of risk, indemnification, and recovery far more frequently than *Folha*. Even as the waste flowed into the Doce, news stories reported on communities down river that sprang into action, rescuing and relocating fish and other aquatic life to lakes and ponds away from the polluted mud flow. *Folha* reported on a "Noah's Ark" task force of 40 people deployed on an improvised emergency mitigation plan to move fish to cleaner waterways. In the immediate aftermath of the disaster, reporters interviewed local fishermen whose source of income and subsistence collapsed along with the dam. Initial detailed accounts of disaster (11 percent) that gave voice to local residents (5 percent) rapidly subsided. One month later, *Estado* cited scientists and government experts as news coverage shifted to discuss the long road ahead to remediation with recovery efforts estimated to last 10 years and costs in the billions.

In July 2016, nine months after the disaster, a special commission assigned to investigate the dam collapse issued an assessment report, prompting both newspapers to pivot from frames of disaster and the risk of contamination to focus on the reasons for the collapse, recovery efforts, and remediation, albeit eschewing investigative or in-depth reporting that might contextualize long-term effects. By September, *Estado* reported the location chosen for the relocation of victims and a federal court called for reparation, damages, and indemnification of victims and the environment. As the news coverage began to frame the disaster as a story of recovery, fewer stories cited victims and their concerns over ongoing risk. Notably, our analysis elicited only one single mention of the impact to the indigenous Krenak peoples whose subsistence revolves around fishing and whose lives and culture were imperiled by the disaster. Rather than report on the populations confronted by a cascade of environmental, socioeconomic, and cultural impacts, within 10 months of the disaster the press increasingly cited court decisions or documents (19 percent), corporate officers and spokespeople (17 percent), government officials (14 percent local and 12 percent national), and their documents and press releases (8 percent).

On the eve of the first anniversary of the dam collapse, Brazil's minister of mines issued the first of numerous statements meant to calm foreign investment markets and to redirect attention to Samarco's role as a major employer and source of tax revenue. In a report from Rio de Janeiro, both newspapers quoted the minister's remark that, "What happened to Samarco was a tragedy, and unfortunate." The emphasis moved away from the impacted communities and victims, although telling, this went unchallenged, as did the government official's headline assertion that local residents wished for Samarco to restart operations. News accounts filed from financial capitals in Brazil and abroad remained uncritical of official sources.

As the focus shifted to recovery and the economic relevance of mining to the state tax base and to the workforce, fewer news stories quoted victims (12 percent), civil society activists, and protestors (5 percent), even as references to "community" rose (N = 16 in 2017). The pivot coincided with the removal in December 2016 of the prosecutor and members of the legal task force assigned to the investigation, a move described by a source in *Estado* as indicative of the power of the mining lobby in Minas Gerais (Werneck, 2016), only two months after the original prosecutor brought charges for homicide and environmental crimes against 21 officials at Samarco, Vale, and BHP Billiton. Despite the power play in the courts, mentions of moral responsibility (18 percent) were outpaced by discussions about remediation (45 percent) and the causes of the disaster seldom surfaced in news accounts (3 percent). Ultimately, discussions focused mostly on a return to mining operations in the valley without mentioning alternative paths to reconstruction or reparations.

There were very few reports about concerns of affected communities, notwithstanding the continued water crisis that prompted more than 300,000 residents in 41 municipalities to rely on bottled water and to abandon subsistence farming over fears of hazardous iron and manganese contamination in underground aquifers. It is telling that coverage of this issue appeared only once in *Folha*, as a report filed from a press junket to a water summit in the Amazon, funded by Coca Cola. The article, which quoted foreign and Brazilian NGO leaders, cited a source critical of the nation's lack of preparedness to recover polluted river basins such as River Doce: "Brazil, which used to be a nation of rivers, is becoming a nation of sewers" (Trindade, 2017).

Amid reports in May 2017 that the Samarco foundation would allocate 1.1 billion reais (US$275 million) over the course of 10 years for reparations and environmental restoration of the river basin, *Estado* again shifted emphasis to stories about recovery and indemnification, whereas *Folha's* increasingly focused on risk. At the close of 2017, two years after the dam collapsed and wrought havoc in the valley and beyond, state officials voted 11 to 1 to grant two of the three licenses needed to return the mining company to operation. *Estado* referred in a news brief item to the loss of 19 lives from the dam break without mentioning the environmental damage, contamination, relocation, recovery, or other impacts on the region.

Conclusions

Local and national news coverage of disasters that inflict environmental, health, and economic harms on communities can provide more than information; they can become constitutive agents and pathways to structure conflicts, define actors and outcomes, and their contents as resources that are disseminated and transformed through other networks (Cottle, 2006; Hutchins & Lester, 2015). How news media define disasters, assign blame and risk, and how actors are depicted and explain events can have significance beyond local and national borders. Yet local and national news provide important points of comparison for understanding nuances in how news media cover disaster and as this chapter considers for those that impact on human, ecological, economic, and political systems.

In the case of the Samarco dam collapse, news reports surveyed here shared similarities. Initial coverage that focused on victims and harm shifted to legal narratives, economic impacts, and a focus on the company, rather than the community. After the initial coverage, the focus became one of official legal, economic, and political angles, with official voices and sources dominating coverage, the latter in accordance with the findings of Mourão and Sturm (2018). Interestingly, and especially in early coverage, both local and national news published more articles from their own reporters, rather than using wire services or international reporting, a shift from other studies examining coverage of environmental disasters far from urban centers (Pinto et al., 2017). Wire news coverage originated mostly from metropolitan centers, possibly due to the mountainous location of the dam, a three-hour drive from Belo Horizonte yet relatively distant from news capitals.

Overall, the regional paper framed more stories in terms of risk and "recovery" in the economic sense, as narratives adhered to questions regarding indemnification and litigation, rather than community resilience, ecological reparation and public health, and well-being. The "tragedy" of the disaster was consistently reframed as an economic one, rather than ecological or communal, and the lack of focus on environmental impacts would be surprising, given the significance for readers, were it not for the prominent role of mining in the economy of the state.

Yet, more than three years after the dam burst, environmental agencies and researchers found that recovery of the ecosystem lagged and that inadequate post-disaster management irreversibly compounded catastrophic impacts in an area where more than one million people's livelihoods depend on the river for drinking and irrigation. Contamination of riparian ecosystems, underground water sources, fisheries, fauna, and vegetation is expected to persist long term, if not permanently, with pollution extending beyond the Atlantic Forest biome, past coastal mangroves, and affecting the Abrolhos coral reefs, the Atlantic Ocean's most distinctive area of biodiversity 200 kilometers offshore (Fernandes et al., 2016). The far-reaching consequences of this disaster has not been grasped by local or national reporters. Local coverage emphasized economic harms and closely followed how official sources have recast the event, rather than telling the stories of the community and seeking to provide a forum for building community resilience.

Shortly after the authors completed this chapter, another Vale tailings dam at Córrego do Feijão (Feijão Stream) collapsed in Minas Gerais, releasing a muddy toxic sludge that left at least 224 people dead (Okumura, 2019). As iron-ore mining waste clouded the Paraopeba River and impacted urban, riverine, and indigenous communities, the headline in *Estado* blared: "Another Crime: Three years after Mariana: dam collapse in Brumadinho causes a new mud tsunami and leaves death and destruction in its wake" (*Estado*, 2019). The news cycle unfolded predictably, with initial episodic focus on loss of life and rescue efforts. Elsewhere in Brazil, nearly all 87 upstream dams share the tailings dam design that potentially hides similar structural flaws (Agência Nacional de Mineração, cited in Darlington et al., 2019). Absent news coverage providing context and the environmental risk associated with extractivist ventures.

This chapter examined a single case study, with a sample of local and national news coverage. Future research should compare various cases in international contexts to provide more understanding of the scalar reach and variables at work in translating and emphasizing various components of "what happened." Further, social media, while not examined here, could provide a window into citizen-produced content about the disaster and its aftermath. But mining ventures remain profit-making industries; news accounts reported BHP, Vale, and other mining stock outperformed giants like Apple, Facebook, and Google in 2018 (Els, 2018). As the exposure of people to disaster risks increase faster than the capacity to reduce vulnerabilities (Freitas et al., 2016), and as global demand for energy, minerals, and other resources targets Latin America (e.g. Casey & Krauss, 2018; Serapio & Xu, 2018), it is imperative that communities receive news that goes beyond short-term, crisis-centered coverage tied to official narratives and toward in-depth, community-oriented discussions that foster long-term resiliency.

References

Abramowitz, M. (2017). Freedom of the press 2017: Press freedom's dark horizon. Freedom House Special Report. Retrieved December 14, 2018, from https://freedomhouse.org/report/freedom-press/freedom-press-2017.

Agurto-Detzel, H., Bianchi, M., Assumpção, M., Schimmel, M., Collaço, B., Ciardelli, C., ... Calhau, J. (2016). The tailings dam failure of 5 November 2015 in SE Brazil and its preceding seismic sequence. *Geophysical Research Letters, 43*(10), 4929–4936.

Branton, R., & Dunaway, J. (2009). Spatial proximity to the U.S. Mexico border and newspaper coverage of immigration issues. *Political Research Quarterly, 62*(2), 289–302.

Carmo, F., Kamino, L., Junior, R., Campos, I., Carmo, F., Silvino, G., ... Pinto, C. (2017). Fundão tailings dam failures: The environment tragedy of the largest technological disaster of Brazilian mining in global context. *Perspectives in Ecology and Conservation, 15*(3), 145–151. https://doi.org/10.1016/j.pecon.2017.06.002

Casey, N., & Krauss, C. (2018, December 24). It doesn't matter if Ecuador can't afford this dam. China still gets paid. *The New York Times.* Retrieved from https://www.nytimes.com/2018/12/24/world/americas/ecuador-china-dam.html.

Cottle, S. (2006). *Mediatized conflict: Developments in media and conflict studies.* Maidenhead, UK: Open University Press.

Darlington, S., Glanz, J., Andreoni, M., Bloch, M., Peçanha, S., Singhvi, A., & Griggs, T. (2019, February 9). A tidal wave of mud. *The New York Times.* Retrieved from https://www.nytimes.com/interactive/2019/02/09/world/americas/brazil-dam-collapse.html.

Els, F. (2018, December 27). BHP, Rio, Vale stock bested Apple, Google, Facebook in 2018. *Mining.com.* Retrieved December 29, 2018, from http://www.mining.com/bhp-rio-vale-stock-bested-apple-google-facebook-2018/.

Estado de Minas. (2019, January 26). Outro crime, três anos depois de Mariana estouro de barragem em Brumadinho provoca novo tsunami de lama e deixa rastros de morte e destruição [Another crime: Three years after Mariana, dam collapse in Brumadinho causes a new mud tsunami and leaves death and destruction in its wake]. *Estado de Minas.* Retrieved from http://impresso.em.com.br/.

Fernandes, G., Goulart, F., Ranieri, B., Coelho, M., Dales, K., Boesche, N., ... Soares-Filho, B. (2016). Deep into the mud: Ecological and socio-economic impacts of the dam breach in Mariana, Brazil. *Natureza & Conservação, 14*(2), 35–45. https://doi.org/10.1016/j.ncon.2016.10.003

Freitas, C. M. D., Silva, M. A. D., & Menezes, F. C. D. (2016). The disaster at the Samarco mining barrage: Exposed fracture of Brazil's limits in disaster risk reduction. *Science and Culture, 68*(3), 25–30.

Funk, M. J., & McCombs, M. (2017). Strangers on a theoretical train: Inter-media agenda setting, community structure, and local news coverage. *Journalism Studies, 18*(7), 845–865.

Gant, C., & Dimmick, J. (2000). Making local news: A holistic analysis of sources, selection criteria, and topics. *Journalism & Mass Communication Quarterly, 77*(3), 628–638.

Gutsche, R. E., Jr., Jacobson, S., Pinto, J., & Michel, C. (2017). Reciprocal (and reductionist?) newswork: An examination of youth involvement in creating local participatory environmental news. *Journalism Practice, 11*(1), 62–79.

Hamilton, J. T. (2004). *All the news that's fit to sell: How the market transforms information into news*. Princeton, NJ: Princeton University Press.

Hatje, V., Pedreira, R. M. A., Rezende, C. A., Schettini, C. A. F., Souza, G. C., Marin, D. C., & Hackspacher, P. C. (2017) The environmental impacts of one of the largest tailing dam failures worldwide. *Scientific Reports 7*. Article number: 10706.

Houston, J. B., Pfefferbaum, B., & Rosenholtz, C. E. (2012). Disaster news: Framing and frame changing in coverage of major US natural disasters, 2000–2010. *Journalism & Mass Communication Quarterly, 89*(4), 606–623.

Hsieh, H. F., & Shannon, S. E. (2005). Three approaches to qualitative content analysis. *Qualitative Health Research, 15*(9), 1277–1288.

Hughes, S., & Lawson, C. (2005). The barriers to media opening in Latin America. *Political Communication, 22*(1), 9–25.

Hutchins, B., & Lester, L. (2015). Theorizing the enactment of mediatized environmental conflict. *International Communication Gazette, 77*(4), 337–358.

Jacobson, S., Pinto, J., Gutsche, Jr., R., & Wilson, A. (2019, in press). Goodbye, Miami? Reporting climate change as a local story. In J. Pinto, R. Gutsche, Jr., & P. Prado (Eds.), *Climate change, media & culture: Critical issues in environmental communication*. To be published by Emerald Publishers.

Meio & Mensagem. (2017). Portfólio de Mídia. Retrieved December 15, 2018, from http://portfoliodemidia.meioemensagem.com.br/portfolio/midia/

Mining Technology. (2018, October 17). Samarco to restart operations with limited capacity in 2020. *Mining Technology online*. Retrieved November 14, 2018, from https://www.mining-technology.com/news/samarco-restart-partial-operations-2020/.

Molotch, H., & Lester, M. (1975). Accidental news: The great oil spill as local occurrence and national event. *American Journal of Sociology, 81*, 235–260.

Mourão, R. R., & Sturm, H. A. (2018). Environmental journalism in Brazil: History, characteristics, and framing of disasters. In B. Takahashi, J. Pinto,

M. Chavez, & M. Vigon (Eds.), *News media coverage of environmental challenges in Latin America and the Caribbean* (pp. 67–90). Cham: Palgrave Macmillan.

Okumura, R. (2019, April 7). Sobe para 224 o número de mortos identificados na tragédia de Brumadinho [The number of dead identified in the Brumadinho tragedy rises to 224]. *O Estado de São Paulo*. Retrieved from https://brasil.estadao.com.br/noticias/geral,sobe-para-224-o-numero-de-mortos-identificados-na-tragedia-de-brumadinho,70002782640.

Phillips, D. (2016, October 15). Samarco dam collapse: One year on from Brazil's worst environmental disaster. *The Guardian*. Retrieved from https://www.theguardian.com/sustainable-business/2016/oct/15/samarco-dam-collapse-brazil-worst-environmental-disaster-bhp-billiton-vale-mining

Pinto, J., Prado, P., & Tirado-Alcaraz, J. A. (2017). *Environmental news in South America: Conflict, crisis and contestation*. Springer.

Pinto, J., & Vigon, M. (2018). Comparing South Florida Spanish-language and Cuban media coverage of sea level rise. In I. B. Takahashi, J. Pinto, M. Chavez, & M. Vigon (Eds.), *Mediating demand, degradation and development: Media reporting of environmental issues in Latin America*. Palgrave Macmillan.

Ploughman, P. (1995). Local newspaper roles in the Love Canal disaster. *Newspaper Research Journal, 16*(2), 56–75.

Quandt, T. (2008). (No) news on the World Wide Web? *Journalism Studies, 9*(5), 717–738.

Ruhfus, J. (2017, February 16). Brazil's river of mud. *Al-Jazeera online*. Retrieved from https://www.aljazeera.com/programmes/peopleandpower/2016/08/brazil-river-mud-160818081002569.html.

Salwen, M. (1995). The power of news, by Michael Schudson. *Newspaper Research Journal, 16*(2), 161–163.

Serapio, M., & Xu, M. (2018, September 18). Vale eyes expansion of Brazil iron ore mine to feed Chinese demand. *Reuters*. https://www.reuters.com/article/us-china-steel-vale/vale-eyes-expansion-of-brazil-iron-ore-mine-to-feed-chinese-demand-idUSKCN1M00AG.

Trindade, E. (2017, December 7). Em meio a crises hídricas e má gestão, país discute agenda de Fórum da Água [Amid hydrolic crises and bad management, the nation discusses Water Forum agenda]. *Folha de São Paulo*.

Waisbord, S. (2000). La politica del palo. *Hemisphere: A Magazine of the Americas, 9*(2).

Werneck, G. (2016, December 7). Após dissolver força tarefa de Mariana, MP garante que acompanhamento vai continuar [After dissolving the Mariana task force, the office of the prosecutor guarantees the issue will continue to monitor]. *Estado de Minas*.

Zhouri, A., Valencio, N., Oliveira, R., Zucarelli, M., Laschefski, K., & Santos, A. F. (2016). The Samarco disaster and the politics of affectation: Classifications and actions that produce social suffering. *Science and Culture, 68*(3), 36–40.

CHAPTER 3

Reporting from the 'Inner Circle': Afno Manche and Commitment to Community in Post-earthquake Nepal

Chindu Sreedharan and Einar Thorsen

We could have been killed that day… so that gave us a new perspective. Till now I am reminded of how I was lucky to survive.

News reporting from areas affected by natural disasters is fraught with difficulties, not to mention dangers, and in recent years, much has been written about the constraints involved in such journalistic endeavours. Both scholars and practitioners, drawing on experiences from sites as varied as Chile, China, Haiti, India, Japan, and Nepal, have highlighted issues that challenge reporters in post-disaster situations: among others, threat to physical safety, informational vacuum, destruction of communication and transport systems, and ethical dilemmas (Puente, Pellegrini, & Grassau, 2013). Considerable debate has also centred around the personal trauma that such intimate encounters with human tragedy inflict on those who bear witness to report (Buchanan & Keats, 2011; Himmelstein & Faithorn,

C. Sreedharan (✉) • E. Thorsen
Bournemouth University, Bournemouth, UK
e-mail: csreedharan@bournemouth.ac.uk; ethorsen@bournemouth.ac.uk

J. Matthews, E. Thorsen (eds.), *Media, Journalism and Disaster Communities*, https://doi.org/10.1007/978-3-030-33712-4_3

2002; Kramp & Weichert, 2014; MacDonald, Hodgins, & Saliba, 2017). These well-documented difficulties acquire a new and more complex dimension when journalists cover 'their own' disasters; when they are forced to chronicle the state of affairs in their own lifeworld. What happens then? How do reporters function when disasters affect their communities? This chapter considers one such instance, the journalism in the aftermath of the 2015 Nepal earthquake, when the professional and personal identities of Nepali news media personnel merged into one—and they became journalists who survived.

Disasters in Nepal

Nepal has a history of severe seismological disturbances spanning centuries and is the 11th most earthquake-prone country in the world. Since 1255, 12 major disasters, with magnitudes ranging up to 8.2 on the Richter scale, have occurred in the country, killing tens of thousands and displacing millions. The death toll in the Great Nepal-India Earthquake of 1934 alone stood at more than 8000. The comparatively weaker tremors of 1980, 1988, and 2011 together killed 849, injured 6965, and damaged, at a conservative estimate, more than 92,414 buildings.

The 7.8 earthquake that occurred four minutes before noon on 25 April 2015, thus, was the latest in an extending series of disasters in the region. It affected 31 of Nepal's 75 districts, 8.1 million people. Together with the 7.3-magnitude aftershock that followed 17 days later, it killed 8794 people and injured 22,300. Nearly 500,000 houses were destroyed and another 288,255 partially damaged, leaving hundreds of thousands homeless to face the upcoming Himalayan winter in temporary shelters. The total economic loss stood at an estimated US $7 billion—more than one-third of Nepal's GDP.

As aftershocks continued, Nepal's disaster preparedness came under revived scrutiny. One of the most damaged countries in the Asia-Pacific region, with a continuing trend of severe deforestation and forest degradation (UNESCO, 2018), Nepal is particularly vulnerable to natural disasters: every year, it sees approximately 500 disaster events such as fire, landslides, floods, and epidemic (Sreedharan, Thorsen, & Sharma, 2019). Despite this, disasters have traditionally been managed on an ad-hoc basis, "attended to as and when they occurred" (Nepal, Khanal, & Sharma, 2018, p. 6). The Natural Disaster Relief Act (1982), the establishment of the Disaster Preparedness Network Nepal (1996), and formulation of the

National Strategy for Disaster Risk Management (2009) and the National Disaster Response Framework (2013) were all steps in strengthening the crisis management capabilities in the country through legislation and policymaking. But Nepal's disaster response was found lacking in the wake of the 2015 earthquake. Although there existed policies to manage disasters, the "framework could not deliver what it seemed to promise," limited, as it were, by both its "structure" and "response capacity" (Manandhar, Varughese, Howitt, & Kelly, 2017).

Nepali Media and the Earthquake

The Nepali news media landscape is characterised by a remarkable number of radio stations. Spread across the country, and particularly serving the rural population, these stations broadcast in a plurality of languages—123 languages are spoken in Nepal, according to the 2011 census—to command substantial reach, higher than other forms of media.

Newspapers are centred in the Kathmandu valley, with low overall print circulation owing to "the difficult geographical terrain, the high recurring costs for both publishers and readers, and the adult literacy rate at only around 60 percent of the population" (Acharya, 2019). Recent years have seen a dramatic increase in television and digital and social media penetration in urban areas, but radio enjoys a far greater reach into rural communities across the nation. Compared to the 189 newspapers published daily and 117 television channels predominantly catering to urban audiences, there are 736 FM radio broadcasters, including 314 community stations. This allows an estimated 98 per cent of the population access to radio (a significant proportion access FM stations via mobile phones).

While the diversity of outlets signals the basis for a robust news media system (a point to note here is that, unlike in many other South Asian countries, radio stations are allowed to broadcast news and news-based programmes in Nepal), it must be remembered that professional media have only had a very short lifespan in the country. Till democratic reform in the late 1990s amidst burgeoning civil war, and the resultant emergence of commercial media houses, journalism was a 'volunteer' career. As such, the profession is still in its nascency in Nepal, facing several training, resourcing, and other developmental challenges. Nepali journalism, hence, was ill-equipped to deal with a disaster of the magnitude of the 2015 earthquakes. Compounding this was the fact that, much like every other aspect of Nepali life, the tremors wrought large-scale destruction to the journalism infrastructure: a 2016 report by the Federation of Nepali Journalists found that 266 media houses were damaged and 1813 journalists directly affected, which included the deaths of 3 journalists and injuries to 14.

Against this backdrop, this chapter explores the experiences of Nepali journalists as members—in a personal and professional capacity—of disaster-affected communities. We draw on in-depth interviews with 46 journalists, editors, and other officials responsible for news dissemination after the earthquake. Interviewees were from the city areas of Kathmandu and Patan, as well as regions worst affected by the earthquake: Sindhupalchok, Nuwakot, and Gorkha. Nearly three-quarters of the journalists and editors interviewed were from radio or newspapers, with the remaining from TV, dedicated news websites, or wire services. Our analysis is also informed by 150 interviews with survivors (including journalists) undertaken for our affiliated post-disaster journalism project, Aftershock Nepal (http://www.aftershocknepal.com).

Afno Manche

Nepal has a collectivist culture, founded on the 'afno manche' family system, where the needs of the family take priority over external obligations. Afno manche is a Nepali term that translates to one's own people, "those who can be approached whenever need arises" (Bista, 1991, p. 98). Similar to the Chinese guanxi network of social relations (Haaland, 2010), afno manche refers to relationships of reciprocity that calls on Nepalese to prioritise the needs of their 'inner circle' first.

Most often these inner circles are knit with kinships, but this need not always be the case. Afno manche can be more strategic in nature as well, to include people unconnected by blood or marriage so as to create "informal personalized organization of activities that affect the operation of formal structures of market and bureaucracy" (Subedi, 2014, p. 56). This practice has drawn criticism for encouraging discrimination on the basis of kinship, caste, and social relations (Jamil & Dangal, 2009) and is often seen as a root cause for the favouritism and nepotism that affect the governmental machinery in Nepal (Subedi, 2014). But such relations, whether deliberately cultivated or natural, run deep in the Nepali society, underpinning the human connectivity in communities there. As Bista (1991) put it, the most important asset for anyone in Nepal is not what you know, but who you know.

Subedi (2014) classifies afno manche networks into four groups: family, business, bureaucratic, and political. As mentioned at the outset, a significant element in all these networks is reciprocity: these relationships are confirmed by practices of symbolic reciprocation such as exchange of gifts, information, and services. The system is dependent on cultural values such as trust, dependability, and loyalty, and unreciprocated requests for help may lead to recalibration in the inner circle. As a university teacher described this obligation: "If one is in a difficult situation but does not get help from afno manche, s/he is no more afno manche" (Subedi, 2014, p. 67).

Of the four networks, not surprisingly the family afno manche is generally the strongest and more durable. Driven by emotions and affection, it is made up of immediate family members, followed by closest extended family—often cohabiting. Both maternal and paternal kinfolk are included in this innermost of the inner circle, as are those treated as family—for instance, classmates (Subedi, 2014). Then comes the other circles of 'insiders' such as friends, neighbours, and people of trust. A Nepali villager exemplified afno manche thus: "[O]ur brothers are closer. In case the brothers are not there then friends are closer, and if friends [are not there] then the neighbours... then comes the Ward Chairman" (Anderson, 2004, p. 6).

The family afno manche, thus, is a network bound by an obligation to help—a strategy of survival at its core, where members close ranks to protect each other from hostilities 'outside,' for the betterment of those within. Anderson (2004, p. 6) describes it as "relations of concentric circles" with "descending obligation or expectation towards the periphery." Though family afno manche may not require immediate or direct quid pro quo, it is this promise that holds the relations in place. When this social norm is broken, when expected obligations are not met in situations of need, members could lose their status within the network and afno manche could become tadako manche (faraway person). Communities we argue are made up of a plethora of afno manche networks, or concentric circles to borrow Anderson's term, that in the case of disasters are mobilised in a spontaneous response to an external threat or event. This is further complicated where the onset of a disaster triggers different responses and needs from an individual's afno manche networks.

Afno Manche and Dual Trauma

The afno manche network came under enormous pressure in the post-disaster days of 2015. The scale of destruction caused unusually severe situations of need in a nation familiar with—but not resilient to—disasters, affecting as it did significantly large segments of population across a geographical spread (Sreedharan et al., 2019). This, as we evidence in the next section, had a significant impact on the post-disaster journalism from Nepal. Not only were Nepali journalists personally affected by the earthquakes, particularly given the concentration of news organisations in the affected Kathmandu region, but many were also affected through their afno manche in regions of harm.

Compounding this *primary* shared trauma of familial networks was the *secondary* trauma brought about by the calls of their profession—which, in times of tragedy, requires their exposure to the suffering of others. Studies so far, mainly from a Western perspective or focused on 'parachute' reporters from outside the disaster community, have found substantial evidence of psychological stress leading to conditions of post-traumatic stress disorder and depression among disaster journalists (Weidmann, Fehm, & Fydrich, 2008; Backholm & Björkqvist, 2012). To withstand these and function better, seasoned reporters adopt several coping strategies: identifying with their professional values, cultivating a strong sense of purpose, creation of personal networks, acquisition of psychodynamic knowledge, and so forth (Himmelstein & Faithorn, 2002). Specific investigations into the experiences of journalists who reported the disasters of their own communities—investigations that focused on their dual trauma as survivors and scribes, specifically—are lacking. But related studies from hazardous situations indicate high levels of mental stress among those who publish from 'within.'

An inquiry into the psychological well-being of Iranian journalists, for instance, found that they experienced extraordinary levels of danger and emotional distress, owing to their inability to separate their work from their broader (and dangerous) social environment (Feinstein, Feinstein, Behari, & Pavisian, 2016). Another study that looked at Kenyan journalists warned of "significant symptoms of emotional difficulties" and the resultant negative impact on "good journalism," which "depends on healthy journalists" (Feinstein, Wanga, & Owen, 2015, p. 1). In a similar vein, a UNESCO-funded study that looked at Mexican journalists found that one in four respondents stopped working on a story because they were too traumatised (Feinstein, 2012). Hughes and Marquez-Ramirez

(2017) build on this body of work. Providing insights into the journalistically limiting risk-reduction strategies Mexican journalists used (among these, self-censorship and avoidance of on-field reporting) to navigate a situation where their professional identities spilled into personal lives, they write: "For many journalists, then, professional practice has clearly become an exercise in balancing personal and occupational risks with perceived duties to society" (Hughes & Marquez-Ramirez, 2017, p. 515).

Here, it is also worth pointing to the evidence that suggests journalists who report on disasters in their geographic locality bring with them a strong(er) sense of commitment to their community, often stepping outside the bounds of traditional journalism. Usher (2009), after a post-Katrina study, writes of how the role of local journalists at *The Times-Picayune* in New Orleans were complicated by their personal experiences and their newspaper became an "advocate for the city." Matthews (2017) describes a "sense of mission" among the local journalists in Ishinomaki after the 2011 Japan tsunami, an obligation to "provide information that would enable people living through the disaster to cope," which extended into the recovery phase (Matthews, 2017, p. 469). With this as context, the following section considers the post-disaster experiences of Nepali journalists.

Reporting Disasters from Within: Experiences of Nepali Journalists

Though many Nepali journalists have experience reporting disasters, the 2015 earthquakes challenged their professional obligations unusually. In the case of the floods, landslides, fire, and the other more localised events they cover every year, they were, more often than not, able to retain their professional identity, functioning not as survivors themselves but as reporters reporting on their own community. The magnitude of the 2015 disaster, however, drew into its ambit a larger cross-section of people, including a significant number of national and regional journalists and their near and extended families.

The Journalists Who Survived

The tension between journalists' professional identity and their emotive response to experiencing the disaster as survivors emerged as a recurring theme. Journalists we interviewed revealed a conflict between coping with their (and their family's) trauma and their perceived professional duty to report. A radio journalist from Sindhupalchok, one of the severely affected areas, spoke of the harrowing time immediately after the earthquake when he was searching for his family members. "After two hours, we get together," he said, "and I am happy that my family, all my relatives are very well." The journalist also spoke of the month afterwards, when aftershocks continued and he had to put aside the calls of his profession, to attend to the safety of his family.

> I lived for a month in a tent. My house had cracks. The kids were very scared and they didn't want to go to the house. The quakes came continuously and I thought, they all thought, they were going to die. I just forgot at the time that I was a reporter.

A magazine editor, who continued working, expanded on this, speaking from personal experience and also drawing from interactions with a team of journalists:

> So we were reporting at such a time, and you know, families are scared, parents are worried, partners for those who have them are worried. But at the same time, it's your job as a reporter, as a journalist, to share the story. But you have to ignore some of that… I had to report back, clock in at work, and clock in at home as well. Had to spend those few hours at home. Because this is when my parents were also staying in tents. So you have to go, they aren't your people, they're worried, they're old… they have to take care of parents who are even older.

An online journalist from Kathmandu remembered the exhaustion he felt in the days after, and the overriding feeling of numbness that seemed to have afflicted all survivors:

> Complete lack of sleep because every five minutes there were aftershocks. Also, everyone had become somewhat numb with the fear of uncertainty. In Nepal there is a term called *satogoye*. It means when someone is in shock or trauma, at that time he becomes numb. So that was the feeling all around. We didn't know what to do.

A newspaper columnist based in the Pathan area of Kathmandu spoke of another aspect of survival and post-disaster journalism as he tried to 'file copy' from home:

> Hotels were getting full and then in my case, my sister's house was in pretty bad [shape], She was living with us and our house became … community house. As I was writing in some of the pieces, your house is not your house anymore. It's a community house, it's a community toilet.

Another experience shared by interviewees was that there was a loss of human personnel at a crucial time from newsrooms in Kathmandu, where many national print and television broadcasters are based. A TV journalist described the situation:

> [M]ost of the reporters were from outside of Kathmandu, so most areas … those reporters have to go back to their house, and we had to help them. We can't say them no, you have to stay here and you have to work.

Many journalists were too affected to work immediately after the earthquake. The disaster also gave some respondents a different outlook. A newspaper journalist in Kathmandu gave voice to this, when he said:

> It's hard to forget the destruction we saw. We couldn't live properly and work properly after that. I couldn't come to my office to write.

These personal pressures impacted the journalism produced by the survivor journalists and many were quick to point this out. A freelance journalist based in Kathmandu said:

> It was very difficult to maintain quality … We just try to collect the information. We were not able to do work properly. There were psychological problems, family obligations … because life comes first before any other things … also need to stay safe, before writing anything.

Another journalist, also from Kathmandu, summarised the balancing of the personal and professional thus: "As a journalist I have a responsibility towards society, but as a son I have my responsibility towards my parents."

Commitment to Disaster Communities

These personal constraints notwithstanding, journalists demonstrated a firm commitment to their community. Many indicated they were aware of the heightened demand for their services post-disaster—to "do some good" and bring "help" to people in a non-routine situation, at a time when "everything is a problem." In line with Matthews's finding in the Japanese context, driven by a strong sense of community responsibility, Nepali journalists embraced the role of "information-disseminator" (Matthews, 2017, p. 475). In this regard, and similar to what Usher (2009) noted in her US study, they appear to have adopted an interventionist role for themselves that moved well beyond the role of the observer they practised in routine news events.

A TV journalist from Kathmandu detailed how he took direct action to help people in need: "We talked to different organisations and arranged food and clothing, and I myself went to give it to people." He also described how he once persuaded aid workers to provide additional support to a victim, using the threat of filming them as leverage to make an intervention. "We thought, ah, at least we did something. And that difference was made by the camera probably," he said.

Journalists described how they were the first to reach remote sites and how they helped rescue efforts by providing information directly to government officials. A newspaper journalist from Gorkha said, "I did not wait to break the news, rather I supplied it to DDRC [Disaster Risk Reduction Portal], so that they could help maximum people in need."

A radio editor from Kathmandu spoke of working directly with survivors, helping the information flow by establishing a makeshift information centre with one table and one or two chairs:

> We didn't think that people would come. After some time we realised, oh, we need some more chairs. People came from the village… so we needed some water also.

One criticism levelled against crisis reporting—in fact, journalism in general—is its gravitation towards accredited sources (Berrington & Jemphrey, 2003; Kim & Lee, 2008), often at the cost of under and misrepresented sections of the afflicted society. An important reason behind this during disasters is first-hand access, or the lack thereof; as also—where external, 'parachute' journalists are involved—an unfamiliarity with cultural contexts and ground realities. Local news media, in some situations, are known to enrich disaster reportage (see Matthews, 2017). Given this context, it was interesting to note that for a section of our interviewees, their commitment to the disaster-affected community extended beyond their immediate afno manche, to include those who would otherwise go unheard.

A Kathmandu-based newspaper journalist reflected on this self-critically, speaking of the Nepali news media's coverage of the day-to-day issues of survivors:

> We have covered so many issues in local media because we were not just taking statements of people in the government or [those] involved in the rescue operation. We have covered the voices of so many unheard. Still we are not able to reach people suffering the most in the remote part of the county, particularly the underprivileged community. They are not getting proper space in media. We have not been able to raise their voices in our reporting—I see lapses in giving proper space to these people.

Going beyond one's family and committing to the disaster community, for some journalists, appear to have also been cathartic. Journalists indicated that this helped them cope with the psychological stress the disaster placed them under, with some describing their element of pride at their decision to keep working. A newspaper journalist from Kathmandu said:

> To be frank, I compromised all my personal obligations. I moved out for reporting. Two days after the earthquake I was the first one to go to the Shorpani, near the epicentre in Gorkha district.

Another journalist, a publisher, described a similar experience and the extremes he went to ensure his journalism continued into the weeks after the disaster. When he could no longer print his own paper, he said, he took up reporting for other news organisations:

> My family suffered and I suffered a lot of loss. In that situation I didn't care about my family... only my work. I just kept working. I lost being able to print my own paper because of the earthquake, that's why I started freelancing and writing for other publications.

National Pride and the 'Outsiders'

The Nepali journalists' solidarity to their community members—to their extended afno manche, as it were—seems to have gone hand-in-hand with a heightened sense of national pride. This feeling of 'us' and 'we-ness' was possibly fuelled, at least to an extent, by the antipathy they developed for a section of international journalists, 'outsiders,' whose conduct many interviewees saw as self-serving and insensitive.

This chimes with evidence from elsewhere. In their comparative analysis of how Korean and US news outlets covered a 1997 air disaster, for example, Kim and Lee (2008, p. 88) cite several instances and crisis studies to make the point that nationalism and ethnocentrism colour international news coverage significantly, and the nationality of the players "is an important predicator of the tone of news coverage" (see also Entman, 2004). Similarly, several scholars who have looked at humanitarian disasters in non-Western contexts make the convincing case that international news media attention is episodic and superficial when it comes to 'distant suffering,' and largely driven by narratives that accentuate the socio-cultural difference between 'us' and 'them' (Joye, 2010; von Engelhardt & Jansz, 2015).

It was not surprising, then, that Nepali respondents indicated the need to close ranks against the 'outsiders.' They spoke much about Nepali resilience and their pride in their local media, often juxtaposing such comments against the unsavoury practices and "mistakes" that foreign reporters made. In essence, international journalists were seen as exploitative, lacking both empathy and understanding of the local context.

A translator exemplified this viewpoint, narrating how a British journalist tried to coerce victims to show more emotion:

> Right after the earthquake, on the second or third day, I started working with a news channel. There was a group of British journalists with celebrity status coming over to Kathmandu. I had a bit of a moral crisis, a sort of an ideological war with them because as soon as they came, I heard them saying that there is nothing award-winning here. When I took them to areas where people were cremating the bodies of family members, that British guy asked me to ask the victims … to display some emotions, to ask them to describe the pain, and that they shouldn't look expressionless. [Here's] a place which has been so devastated, people are already living in so much trauma, turmoil and crisis… and then a journalist comes and all he needs is an award-winning story.

Respondents also said international journalists chased the high-profile Mount Everest region, ignoring the lesser-known but more affected areas in other districts. They had the "wrong mindset" to report the earthquake, a reporter said, as they were only interested in the suffering. Another interviewee captured the disdain many felt for the international counterparts, noting that one journalist "had to leave because their princess was giving birth to a baby, and it was more important for them to cover this news."

A print journalist, who worked for a Nepali daily as sub-editor-cum-reporter, commented on the lack of knowledge and preparedness that some international reporters exhibited:

> At the time of the earthquake, they come and say Kathmandu is this, this, this, this. But the people, their reporting, it was not proper and lots of mistakes! If I am going to any country in the world, I look at the country, the map, all the details—the population, parliament, all information... But they come and say, look, this is final. But this is not final, that's the problem. They were making mistakes with the geography, information also—lots of mistakes.

Nepali journalists felt particularly let down by Indian journalists, who were criticised for both hijacking the narrative to present a Delhi-centric view (Gyawali, 2015) and their insensitive newsgathering approaches. Some Indian broadcast reporters were seen poking injured people stuck under debris with their microphones to elicit a response, Regmi (2016) notes. He goes on to say: "Their acts of news reporting and visualization went beyond a general practice of a true and ethical journalism" (Regmi, 2016, p. 89).

Our interviewees added to this body of criticism, particularly pointing to the conduct of their Indian counterparts in the recovery phase when India imposed an unofficial economic blockade on Nepal. An English newspaper journalist referred to the coverage of the Indian media in the relief phase as "irritating ... one of the worst forms" of journalism. Another interviewee expanded: "They come down, they want this case where a woman had been recently widowed or children has died."

Respondents also pointed out that the national interests of the Indian journalists biased the way they reported the economic blockade and its impact on a nation struggling to recover from a disaster. A Kathmandu-based woman journalist captured this sentiment well:

> Most of the journalists live in Delhi and they have good relationship with the Indian establishment and the people living in India. So they tried to analyse Nepal through the lens of India ... India imposed blockade, disagreeing with some provisions of the Nepali Constitution. Many Indian journalist came to see the problems faced by the Nepalese people because it was post-disaster scenario. But they didn't give enough coverage to the place of India in the blockade so that the earthquake victim could get relief. If they had reservation about the Constitution, they can ask some senior political leaders. But they should speak against blockade that caused serious humanitarian crisis in the country.

Outsider journalists, both Indian and others, also came under criticism for their fleeting reportage and presence in Nepal. As a freelance feature journalist put it, they focused on "damage, damage" and nothing beyond. "They did not care about the reconstruction and rehabilitation of Nepal," another journalist said. "The Nepalese people had to wait till the country marked the first anniversary for that."

In contrast to this, the majority of the interviewees were complimentary of the way Nepali media performed. As evidenced earlier, they functioned under a particularly difficult situation, facing multiple constraints—but though self-critical and quick to point out their limitations, they took pride in both their professionalism and commitment as well as the coverage they provided in general. A journalist expressed the differences in coverage thus: "Nepali media gave what was happening. International media, particularly Indian media, was provocative." Another interviewee stressed the commitment of her fellow journalists, saying:

> I would say national media did a clean job. Radio was the most effective tool. Radio stations collapsed but the people were providing the information. Taking shelter under the tent, they were providing information round the clock.

Survivor Journalism

The challenges local journalists face in disaster situations are often overlooked. So are the important contributions they make to post-disaster journalism. By analysing the experiences of Nepali national and regional journalists after the 2015 earthquake to understand how they responded to a disaster that engulfed their communities, this chapter has highlighted several important insights for our understanding of this complex relationship.

After the earthquake, Nepali journalists entered a grey area in terms of their professional identity. Such was the scope of the disaster that a significant number of journalists were affected. In fact, everyone we interviewed was a survivor, some affected more than others. Their dual status created a clash between their personal and professional selves, an identity dissonance, that had to be constantly negotiated, placing them under additional stress.

This status also exposed them to dual trauma, arising from two quarters. On the one hand, they had to deal with the distress of being a survivor, either living through the destruction themselves or being affected through their afno manche. On the other, functioning as they did as news personnel (even as they mentally mediated their personal and professional responsibilities), they were exposed to the secondary trauma that all disaster journalists face when reporting human suffering. Further, the exposure of the Nepali journalists to the disaster was particularly *prolonged*: unlike their international counterparts who had the relief of removing themselves from the situation, Nepali news personnel continued—still continue—to produce 'survivor journalism' from their post-disaster community.

It is only natural, then, that all this influenced their journalistic outputs. The quality of their reportage, as interviewees indicated, was compromised by the stress and worries about the well-being of their afno manche, not to mention the logistical, infrastructural, and editorial issues post-earthquake Nepal faced (see Sreedharan & Thorsen, 2018; Sreedharan et al., 2019). At the same time, though, Nepali news personnel displayed commendable journalistic resilience, and, echoing earlier studies (Matthews, 2017; Usher, 2009), a strong, interventionist commitment to their communities. National and regional journalists were, thus, eager to make use of their professional status to do what they could for those in need.

It is also conceivable their personal experiences influenced not just their outputs, but their *outlook* of disaster journalism, including their expectations from the news media in general and the international media in particular. Nepali journalists emphasised the need for empathetic reportage, responsible, and sensitive to their cultural context—one which went beyond the "damage" narrative and crossed over into the recovery phase to discuss underlying issues in a sustained manner. Their sharp criticism of the international media and their very evident sense of pride in their own work at the local levels, particularly via community radio stations, could be seen in this context—not only as evidence of their 'we-ness' but as insights from insiders on what is required (and missing) in post-disaster journalism.

In this context, it is worthwhile to underline the need for supporting what we term survivor journalism. At a time when non-routine disasters routinely percolate news bulletins, the reportage of own disasters from within is crucial in the rebuilding of post-disaster societies. So far, though, scholarly and professional attention has largely been monopolised by the newswork of international journalists. It is hence important that a shift is brought about to acknowledge the distant suffering of survivor journalists and to develop survivor journalism.

References

Acharya, U. (2019). Nepal—Media landscapes. *European Journalism Centre*. Retrieved July 1, 2019, from https://medialandscapes.org/country/nepal.

Anderson, H. G. (2004). *What a village can tell—Democratization in Nepal*. Paper for Panel 21 Presented at 17th European Conference on Modern South Asia Studies, Lund, Sweden, 6–9 July 2004.

Backholm, K., & Björkqvist, K. (2012). Journalists' emotional reactions after working with the Jokela school shooting incident. *Media, War & Conflict, 5*(2), 175–190.

Berrington, E., & Jemphrey, A. (2003). Pressures on the press: Reflections on reporting tragedy. *Journalism, 4*(2), 225–248.

Bista, D. B. (1991). *Fatalism and development: Nepal's struggle for modernization*. Delhi: Orient Longman.

Buchanan, M., & Keats, P. (2011). Coping with traumatic stress in journalism: A critical ethnographic study. *International Journal of Psychology, 46*(2), 127–135. https://doi.org/10.1080/00207594.2010.532799

Entman, R. (2004). *Projections of power: Framing news, public opinion, and US foreign policy*. Chicago: University of Chicago Press.

Feinstein, A. (2012). Mexican journalists: An investigation into their emotional health. *Journal of Traumatic Stress, 25*(4), 480–483.

Feinstein, A., Feinstein, S., Behari, M., & Pavisian, B. (2016). The psychological wellbeing of Iranian journalists: A descriptive study. *Journal of the Royal Society of Medicine Open, 7*(12), 1–5. https://doi.org/10.1177/2054270416675560

Feinstein, A., Wanga, J., & Owen, J. (2015). The psychological effects of reporting extreme violence: A study of Kenyan journalists. *Journal of the Royal Society of Medicine Open, 6*(9), 1–6. https://doi.org/10.1177/2054270415602828

Gyawali, S. (2015). A critique of Indian media's coverage of the Nepal earthquake. *Asian Politics & Policy, 7*(4), 664–666.

Haaland, G. (2010). Reflection on contrasting views on themes in Chinese civilization. *Dhaulagiri Journal of Sociology and Anthropology, 4*, 1–20.

Himmelstein, H., & Faithorn, E. P. (2002). Eyewitness to disaster: How journalists cope with the psychological stress inherent in reporting traumatic events. *Journalism Studies, 3*(4), 537–555. https://doi.org/10.1080/14616700220000019173

Hughes, S., & Marquez-Ramirez, M. (2017). Examining the practices that Mexican journalists employ to reduce risk in a context of violence. *International Journal of Communication, 11*, 499–521.

Jamil, I., & Dangal, R. (2009). The state of bureaucratic representativeness and administrative culture in Nepal. *Contemporary South Asia, 17*(2), 193–211. https://doi.org/10.1080/09584930802346497

Joye, S. (2010). News discourses on distant suffering: A Critical Discourse Analysis of the 2003 SARS outbreak. *Discourse & Society, 21*(5), 586–601.

Kim, H. S., & Lee, S. (2008). National interest, selective sourcing and attribution in air disaster reporting. *Journal of International Communication, 14*(1), 85–103. https://doi.org/10.1080/13216597.2008.9674723

Kramp, L., & Weichert, S. (2014). Covering the world in despair: A survey of German crisis reporters. *Journal of War & Culture Studies, 7*(1), 18–35.

MacDonald, J. B., Hodgins, G., & Saliba, A. J. (2017). Trauma exposure in journalists: A systematic literature review. Fusion, no. 11. Retrieved July 1, 2019, from https://researchoutput.csu.edu.au/ws/portalfiles/portal/21892336/19354278_Published_article_OA.pdf.

Manandhar, M. D, Varughese, G., Howitt, A. M, & Kelly, E. (2017). Disaster preparedness and response during political transition in Nepal: Assessing civil and military roles in the aftermath of the 2015 earthquakes. *The Asia Foundation*. Retrieved from https://asiafoundation.org/wp-content/uploads/2017/04/Disaster-Preparedness-and-Response-During-Political-Transition-in-Nepal.pdf.

Matthews, J. (2017). The role of a local newspaper after disaster: An intrinsic case study of Ishinomaki, Japan. *Asian Journal of Communication, 27*(5), 464–479. https://doi.org/10.1080/01292986.2017.1280065

Nepal, P., Khanal, N. R., & Sharma, B. P. P. (2018). Policies and institutions for disaster risk management in Nepal: A review. *The Geographical Journal of Nepal, 11*, 1–24.

Puente, S., Pellegrini, S., & Grassau, D. (2013). Journalistic challenges in television coverage of disasters: Lessons from the February 27, 2010, earthquake in Chile. *Communication & Society/Comunicación y Sociedad, 26*(4), 103–125.

Regmi, K. D. (2016). The political economy of 2015 Nepal earthquake: Some critical reflections. *Asian Geographer, 33*(2), 77–96. https://doi.org/10.1080/10225706.2016.1235053

Sreedharan, C., & Thorsen, E. (2018). Voices from Nepal: Lessons in post-disaster journalism. Centre for the Study of Journalism, Culture and Community. Retrieved July 1, 2019, from http://www.aftershocknepal.com/report/.

Sreedharan, C., Thorsen, E., & Sharma, N. (2019). *Disaster journalism: Building media resilience in Nepal.* UNESCO Kathmandu and Bournemouth University. Retrieved October 1, 2019, from https://www.aftershocknepal.com/disaster-journalism-book/.

Subedi, M. S. (2014). Afno Manchhe: Unequal access to public resources and institutions in Nepal. *Dhaulagiri Journal of Sociology and Anthropology, 8*, 55–86. https://doi.org/10.3126/dsaj.v8i0.10722

UNESCO. (2018). *Getting the message across reporting on climate change and sustainable development in Asia and the Pacific: A handbook for journalists.* Retrieved July 1, 2019, from https://en.unesco.org/sites/default/files/getting_the_message_across_climate_change_asia_pacific_web_2018.pdf.

Usher, N. (2009). Recovery from disaster: How journalists at the New Orleans Times-Picayune understand the role of a post-Katrina newspaper. *Journalism Practice, 3*, 216–232.

von Engelhardt, J., & Jansz, J. (2015). Distant suffering and the mediation of humanitarian disaster. In R. E. Anderson (Ed.), *World suffering and quality of life* (pp. 75–83). New York: Springer.

Weidmann, A., Fehm, L., & Fydrich, T. (2008). Covering the tsunami disaster: Subsequent post-traumatic and depressive symptoms and associated social factors. *Stress and Health, 24*, 129–135. https://doi.org/10.1002/smi.1168

CHAPTER 4

Kesennuma's *Building for the Future* and Ishinomaki's *Rolling Press*: Sharing Localised News of Recovery from Tōhoku's Disaster-affected Communities

Jamie Matthews

The earthquake and tsunami that struck Japan in March 2011 was an unprecedented disaster that destroyed coastal communities in Tōhoku and led to the deaths of more than 22,000 people (Japan Fire and Disaster Management Agency, 2016). It triggered a second event, the nuclear emergency at Fukushima and the subsequent evacuation of over 100,000 residents from the towns and villages in proximity to the Daiichi plant.

While generally evidence shows that international news media focused on the nuclear emergency at Fukushima, with the tsunami and its impacts on coastal communities secondary to global concerns over the effects of a prolonged release of radioactive material (Matthews, 2019; Pantti, Wahl-Jorgensen, & Cottle, 2012), the devastation wrought upon Tōhoku by the tsunami still drew extensive news coverage.

J. Matthews (✉)
Bournemouth University, Bournemouth, UK
e-mail: jmatthews@bournemouth.ac.uk

J. Matthews, E. Thorsen (eds.), *Media, Journalism and Disaster Communities*, https://doi.org/10.1007/978-3-030-33712-4_4

The port cities of Ishinomaki and Kesennuma in Miyagi prefecture were amongst the areas hardest hit by the tsunami. Their topography and relative size meant that both suffered widespread destruction. Ishinomaki, Miyagi's second city, with many lower-lying residential areas, faced the highest death toll from the disaster, with 3890 people listed as killed or missing due to the tsunami (Japan Fire and Disaster Management Agency, 2016). The entrance to the narrow bay upon which Kesennuma sits amplified the height of the tsunami waves and the city also suffered extensive damage, in particular to its waterfront. In the days following the tsunami, Kesennuma also experienced a number of major fires as fuel and chemicals leaked from storage tanks and ignited debris swept away by the waves. The city also lost many of its citizens in the disaster, with over a thousand people listed as killed or missing (Japan Fire and Disaster Management Agency, 2016).

The region's commercial fishing industry, vital to the local economies of Ishinomaki and Kesennuma, was decimated by the disaster, with many fishing boats and the infrastructure upon which the industry relies destroyed by the tsunami.

Both cities featured prominently in the international news coverage of the disaster. Many striking images of the tsunami and its impacts, such as video footage captured of waves sweeping into built-up areas and pictures of a stranded trawler and a young woman wrapped in a blanket amidst the debris were from these two communities. They also featured in the stories of loss and survival that emerged in the aftermath of the disaster. The most significant being the tragedy that occurred at Okawa Elementary in Ishinomaki where more than 80 pupils and teachers lost their lives after the tsunami swept through the school's playground.

Beyond this immediate impact phase, however, far less information has made its way outside of Japan about these communities as they have rebuilt and attempted to recover from the disaster. This is a characteristic of disaster journalism, where news reporting details disaster effects but the issues raised by recovery and reconstruction are seldom considered newsworthy for audiences beyond the affected areas (Houston, Pfefferbaum, & Rosenholtz, 2012; Pantti et al., 2012; Ploughman, 1995). Therefore, while we may learn about the difficulties that communities face following disaster, we hear very little about their longer-term effects and the process of recovery. International media coverage of the Japan disaster generally reflected this pattern. On the first anniversary of the disaster, memorialised coverage looked back at events and returned to some of the communities affected by the tsunami to assess the state of recovery (for example, see Harlan, 2012). This was disproportionate, however, to the attention afforded to the Tōhoku region in the days after the tsunami and as fears grew over the emerging situation at Fukushima. While the issues associated with post-disaster recovery in Tōhoku have understandably remained salient for media, public and policy agendas in Japan, they have received scant attention from media outside of Japan.

This chapter evaluates two grassroot and non-profit media initiatives, the *Rolling Press*, established by a small group of volunteers living in Ishinomaki City, and Kesennuma's *Building for the Future Project*, instigated after the 2011 tsunami that have sought to publish, curate and share localised news and information in English about these two recovering communities. Both initiatives, this chapter will argue, are indicative of the diverse forms of grassroots media that may emerge in post-disaster contexts and contribute to recovery processes. Drawing on interviews with contributors to these two projects, and by highlighting examples of the content that they produced, share and continue publish, it will explore how they have attempted to reach audiences that are geographically dispersed (Reader & Hatcher, 2012), in particular by harnessing the opportunities afforded by digital media. It will also reflect on some of the difficulties that the Rolling Press faced in attempting to achieve longer-term sustainability. These two cases, the chapter will argue, demonstrate how different notions of community may emerge in the context of post-disaster recovery and the potential for grassroots media projects to meet their information needs, illustrated by an attempt to provide news for a broader community of concern that is distant from but shares a connection to those directly affected by the disaster.

Ishinomaki's Rolling Press and Kesennuma's Building for the Future

The two projects that provide the focus for this chapter share similar characteristics but also diverge in other respects, most significantly in the medium they adopted to reach their audiences. It is to these features that discussion will now turn, before addressing the overarching research aims for the chapter.

The Rolling Press was the brainchild of a small business owner in Ishinomaki, who together with two other contributors started the project in 2012 with the explicit aim of making news and information about

Ishinomaki and its recovery from the disaster available in English. The project centred on the production of a small, 11- to 15-page, quarterly colour magazine. The first edition of this was published in September 2012 and alongside translations into English of news items from other outlets it also featured original contributions from local people, all of whom were amateur journalists. This first issue also included an article about how tsunami debris was being processed in the city, another that introduced prominent landmarks in and around Ishinomaki, and some shorter news in brief items about the opening of a market hall in the city and forthcoming events in the city. It also included an English translation of a longer feature article that was originally published in the regional *Sanriku Kahoku* newspaper that presented a local person's experience of the disaster. This piece titled, *My Story of 3/11*, became a long-running feature series in the magazine.

Contributors to the project funded a small print run of the first edition of the magazine but to cover further costs for production, printing and distribution the project adopted a subscription model, requiring subscribers to pay a small annual fee to receive four issues of the magazine. As discussed further below, initially the intended readership for the magazine included non-Japanese volunteers who had travelled to Ishinomaki to assist after the tsunami and had requested up-to-date information about the area and its recovery. By October 2013, following the publication of the fourth issue of the magazine, over 1700 copies of the magazine had been distributed to its subscriber base (NHK, 2013).

Contributors continued to support the quarterly publication of the Rolling Press magazine through 2013 and into 2014, with issues produced during this period using the same colour format to provide localised news and information about Ishinomaki. The seventh issue published in September 2014, three and half years on from the disaster, for example, included an original piece about a mobile application that would enable visitors to visualise the extent of tsunami damage to Ishinomaki and another that presented reflections from disaster volunteers that had now settled in the area. By this time, a number of major reconstruction projects had also been approved or started in the city, including the development of a new municipal hospital and public housing complex. Brief details about these schemes, for instance, and their progress were provided in issues five and six of the magazine.

Quarterly issues of the Rolling Press continued to be published through until the end of 2014, with the eighth issue of the magazine being the last to be sent to subscribers. As the evaluation of the two cases below indicate, the project was initially intended to produce a limited run of the magazine, with the founding editor recognising that the project would have only been able to continue with external support. The Rolling Press established a Facebook page to raise awareness of the project, provide information about subscription options and announce the publication of new issues of the magazine. The page also included pictures of magazine covers and short summaries of some articles carried in later issues. It did not provide links to individual articles and was used primarily for promotional purposes.

The Kesennuma Project launched a Facebook page (www.facebook.com/kesennuma) on 8 August 2011 to share information, news and content about the city aimed at an English-speaking audience. The page was titled *Kesennuma Building for the Future*, a name, according to one of its contributors, that represents the city's aim to 'have a positive recovery from the disaster.' At the time of writing, the page continues to share up to three posts each week.

Like the Rolling Press, the seeds of an idea were sown by one person, a former native of Kesnnuma now living in Tokyo, who wanted to help the city in its recovery from the disaster. The Kesennuma project, as discussed further below, leveraged the opportunities offered by social media to connect with like-minded individuals and identify potential contributors with relevant skills, in particular those bilingual in Japanese and English, and experience who would be able to administer, identify, produce and translate content for the page. In contrast to the Rolling Press, which was written and put together almost exclusively by people local to Ishinomaki, longer-term residents and those who had moved recently to the area, the Kesennuma project was brought to fruition by a small team that were geographically dispersed. Some contributors were living in Kesennuma but others, although having connections to the city, were from further afield, including the current editor of the page who lives in the United States.

The Kesennuma page has continued to post and share a range of different locally relevant content about the city and at times the wider area, Miyagi prefecture and also the Tōhoku region. Posts to the page, similar to the articles carried by the Rolling Press, include sharing and translating into English news and features identified from other media outlets that have a local angle or relevance. This may be linking to a piece about Kesennuma published online by an English-language news provider, such as the *Japan Times*, or providing translations of articles published in local newspapers, for example pieces originally published in the *Sanriku Shimpo*, a supplement to the regional block newspaper of Tōhoku, the *Kahoku Shimpo*. Such articles are able to offer insights into reconstruction and disaster-recovery projects in Kesennuma but also reflect the breadth of community-focused stories and news covered by the local press. Some posts to the page are simply curated and shared from other sources, all however offer a short summary of each item and discuss its relevance for readers seeking news about Kesennuma.

The Kesennuma page also includes original posts and content from its contributors. Across the lifecycle of the project, these have covered a range of topics and issues, including updates on reconstruction efforts in the city, informative posts about the local area, traditions, festivals and how they are celebrated in Kesennuma or highlighting achievements of local people. The page also reflects the breadth of content that can be shared on digital platforms, with videos, photos and links to blogs posted to the page. This multimedia content is drawn from a range of different sources. Pictures, for example, are sometimes provided by contributors to the project or are sourced from the city authority's public relations team.

The Kesennuma page continues to provide information and share news and stories from the community. At the time of writing, the most recent post was announcing the imminent opening of new sections of a major road that would connect Kesennuma directly to the region's major city, Sendai. As of February 2019, the page had 3782 total follows and 3808 likes, indicating a small but sustained and engaged audience for this localised news and content about Kesennuma.

Motivations, Sustainability and Supporting Recovery

Interviews were conducted either face to face or via email with selected contributors to these two projects, including both editors. The evaluation presented here draws out the principal themes emerging from the interviews and, where relevant, the discussion is illuminated through reference to examples of the content produced and shared by these grassroot initiatives.

The overarching rationale for establishing both projects was to reach audiences with localised news and information in English about these two recovering communities. The interviews indicate, however, a broader set of objectives for the Kesennuma project, with the project's founder describing how the page was set up to meet three aims. First, it sought to raise global awareness of Kesennuma; second, to help Kesennuma become internationalised; and, third, to facilitate the exchange of information and support between Kesennuma and people living outside of Japan. In contrast, the editor of the Rolling Press emphasised that the project was started to 'spread news in English' and in response to requests that they received from non-Japanese-speaking volunteers who that had travelled to Ishinomaki after the tsunami and sought updates and news of the city's recovery. They explained how they had become aware that the US news media occasionally carried images and video footage showing Ishinomaki in the months after the disaster but that these reports did not provide any details about the progress of recovery and reconstruction in the city.

The connection with volunteers, those who travelled to Tōhoku after the disaster, is a recurring theme emerging from the interviews. In the six months after the tsunami, over one million volunteers came to Tōhoku to help clear up debris and provide support to affected communities, such as running soup kitchens, delivering food supplies and renovating damaged houses (Zenshakyo, 2013 cited in McMorran, 2017). The vast majority were Japanese nationals but there were also a significant number of international volunteers, expats living in Japan and those from overseas, who came to provide assistance.

Wanting to reach those who had established relationships with the community through their volunteering efforts was a clear incentive for starting the Rolling Press. The founding editor recounted how another contributor had encouraged them to start the project, telling them it was 'their duty' to give something back to those who had supported the city after the tsunami.

Other contributors to the Kesennuma project identified additional objectives and motivating factors, those that may also reflect their own vision for the initiative and reasons for participating. These included describing the aims of the project as 'to help people remember,' after the international media has moved on from the story, 'share the status of recovery to a worldwide readership' and 'attract people that do not know about Kesnnuma.'

The broader ambitions of the Kesennuma project are indicative of the potential for grassroots media to be part of the fabric of disaster recovery, one that is community orientated. Its stated aims were to provide access to news and information about the city. The project also, however, sought to contribute to disaster recovery by raising awareness of Kesennuma and encouraging people to visit the area. This is reflected in some of the content that is shared on and produced for the page, which also features information about the city and region that would be of interest to visitors, such as posting pictures of the natural environment and information about cultural festivals.

These profile-raising ambitions were also consolidated by establishing closer working relationship with other local organisations, in particular the city authority. This is illustrated by some of the content shared on the Kesennuma page, which includes translations of posts to the city's Facebook page, *Hamarainya Kesennuma* (www.facebook.com/prkesennuma), which is maintained by the city authority's PR team, and the use of photos and videos attributed to *Kesennuma Public Relations*. As one interviewee explained:

> We post photos of construction work provided by the city's public relations team, which keep readers informed of where we are on the long road of disaster recovery. New apartment buildings for tsunami survivors and a new bridge over to Oshima Island are good examples… We try to post content of interest to readers by introducing Kesennuma's culture, food, and beautiful scenery as much as we can.

Despite this partnership with the city authority, and the use of its content, interviewees were keen to stress that the administration, sharing and authoring of content for the Kesennuma page was the work of the project team, all of whom gave their time without compensation.

In contrast, the founding editor of the Rolling Press suggested that, although the project was welcomed warmly by many people in Ishinomaki, it was unable to receive formal support from the city.

Both projects were realised by bringing together a group of like-minded volunteers. The founding editor of the Rolling Press acknowledged that while they had no previous experience in journalism or publishing, or indeed felt confident in translating into or writing in English, they were able to find other people living locally that could contribute to the project, by translating articles, copy-editing and helping with the design and layout of the magazine. The first issue was put together by just three people but later issues featured other contributors that authored original articles in Japanese or helped to translate articles from Japanese into English. All the supporters of the Rolling Press project were described as 'amateurs,' without journalistic expertise or experience.

The founder of the Kesennuma project explained how they had used social media, Facebook in particular, for introductions to people who could bring to fruition the idea of an English online news and information site. Through these contacts they identified people who had connections to Kesennuma and would be able to use their skills and experience in writing content, translating, editing and in digital design to contribute to the project. Three contributors, including the current editor of the page who was described by another participant as a writer with 'rich experience in editorial,' live in the United States, but all have associations with Kesennuma. Other contributors are based in Japan, including another Kesnnuma native, now living in Tokyo, with a design background who was able to create banners and icons for the Facebook page and also offline content, such as posters and brochures, that could be used to promote the page. A number of translators also support the project, translating content and articles from Japanese to English and monitoring Japanese and international media for content about Kesennuma that can be shared to the page.

The Kesennuma project, therefore, not only used social media as a way to share locally relevant news and content but to seek contributors and facilitate engagement in the initiative. Although the Rolling Press later set up an online site and Facebook page, this was primarily to promote the magazine. The editor of the Rolling Press acknowledged that for the project to have continued it would have been necessary to move towards a digital or online version of the magazine. Ultimately, sustainability would have required a greater commitment than its small team of contributors were able to give to the project, and the eighth issue of the Rolling Press, published in October 2014, was the last to be produced and distributed to subscribers.

The Kesennuma project, however, continues to publish and share content to its *Building for the Future* page. The ease of using Facebook as a platform to reach audiences to some degree explains the success of the project in being able to continue to deliver relevant content about this recovering community. Providing a small number of posts each week, whether that is sharing and translating stories from other media or authoring original articles, is easier for a small group of volunteers to manage alongside other responsibilities than producing a complete, albeit small, magazine every three months. This limitation was something that was acknowledged by the founding editor of the Rolling Press when they described how alongside running the project they also had to rebuild their own business after the disaster. A Facebook page, or a digital version of magazine, negates the time and logistics needed to produce and send hard copies of a magazine. These factors, combined with the experience in journalism and design of many of the contributors to the Kesennuma page and the support obtained for the initiative from the local community, also explain why, at the time of writing, it continues as a valuable source for those seeking news about Kesennuma as it recovers from the disaster.

Finally, it is important to note that the Rolling Press, and its efforts to bring news of recovery and reconstruction to audiences outside of Japan, came to the attention of journalists in 2013, with a number of items appearing in the international media about the initiative. An extended report about the Rolling Press aired on NHK World in October 2013 that introduced the project and featured interviews with its contributors. Another article, published on the second anniversary of the disaster in March 2013, reported on recovery in Tōhoku and underlined the slow pace of reconstruction in Ishinomaki. It also explained how the Rolling Press, as a community media initiative, was seeking to highlight the realities of disaster recovery for Tōhoku's disaster-affected communities (Adelstein & Stucky, 2013).

Further Discussion and Conclusion

The two cases evaluated in this chapter demonstrate the continued issue of sustainability for grassroots media projects. For the Rolling Press the difficulties in financing the project longer term and the needs of contributors to balance their commitment to the project with the practical concerns of disaster recovery were cited as the primary reasons for discontinuing the project in 2014.

The use of an existing corporate media platform, Facebook, by the Kesennuma initiative, which facilitated the organisation and running of the project and also allowed the curation and sharing of content published elsewhere, reduced considerably the start-up investment required to reach potential audiences. Moreover, identifying and recruiting contributors with relevant professional experience, in particular in journalism and editing, was beneficial for the project.

In many ways, the Kesennuma page operates as a small online hyperlocal news operation, without a need to generate income and is able to operate due to the involvement of 'civic-minded' volunteers (Kurpius, Metzgar, & Rowley, 2010). The short-term sustainability of the Kesennuma project was achieved by leveraging the benefits of this digital media platform and using it to distribute news and content (Van de Wurff, 2002 cited in Kurpius et al., 2010).

While later in the lifecycle of the Rolling Press, steps were taken to move towards the digital distribution of news and information about Ishinomaki and its recovery, including the production of a pdf version of the magazine and the development of a website for promotional purposes, the original model was subscription to a print copy of the magazine. It is inevitable therefore given the project's emphasis on producing original content about the city and its recovery that in the longer term it would have required external funding in some form to continue (Kurpius et al., 2010).

The evaluation suggests that while both initiatives emphasised the principle of seeking to support recovery in their respective communities, the Rolling Press was founded upon and articulated a stronger advocacy approach. This emerged as theme from the interview with the founding editor but was also evident through the types of stories that were presented in the magazine, with some pieces, in particular those focusing on rebuilding and recovery in Ishinomaki, criticising the policies and priorities for disaster recovery in Tōhoku, for example identifying a failure to recognise the needs of particular groups and highlighting concerns about the speed of reconstruction. This theme was also evident in coverage in the international media of the Rolling Press project, which noted that its aims were to share information about the situation in the disaster area to people living overseas and that in part this was due to mainstream media in Japan neglecting recovery from the tsunami (for example, see Adelstein & Stucky, 2013).

As the evaluation of these cases suggest, by raising issues and reporting on recovery in Ishinomaki the Rolling Press adopted a more active role in post-disaster recovery. The Kesennuma project, conversely, sought to contribute to recovery in other ways, most obvious from the content shared to the page is to encourage people to visit the city and Japan's disaster-affected region. Both forms of support, highlighting issues associated with post-disaster recovery and seeking to improve the economic and social conditions for communities rebuilding from disaster, for example by encouraging tourism and inward investment, are important to community-level recovery (Mannakkara, Wilkinson, Willie, & Heather, 2018).

These two cases also demonstrate how different notions of community are relevant in the context of post-disaster recovery and the role of other media forms in meeting these communities' information needs. The priorities for disaster recovery should always centre on the immediate affected community: those who have lived through and whose lives have been forever changed by disaster. The Rolling Press and Kesennuma initiatives, however, also demonstrate the existence and need to cater for a broader community of concern, one that may be geographically dispersed but share a connection to these disaster-affected cities. Both these projects, therefore, sought to address the scarcity of news about recovery that is accessible to non-Japanese speakers and recognised that the disaster, through the extent of coverage in global news and the numbers of volunteers that travelled to the region after the tsunami, had expanded this community of concern. It can be argued then that the disaster and how people may have responded to the plight of affected communities in Japan may have fostered a sense of connection to particular towns or cities devastated by the tsunami, in addition to those who already had established ties with Japan's disaster-affected region. Moreover, the localised content made available to their subscribers, readers and followers by the Rolling Press and Kesennuma projects may have served to consolidate connections with these two recovering communities. In this sense both projects demonstrate the essence of contemporary community media by articulating, supporting and providing for a community that is born out of human connectivity and one that is not confined to geographic boundaries (Robinson, 2014).

The cases illuminated in this chapter imply that it is necessary to recognise these secondary interest-based communities, those that may not be directly at risk from or impacted by disaster events but share a connection, concern or seek to support those affected by disaster or trauma. If we accept the globalised nature of disasters, where disasters activate concerns and responses from broader publics (Cottle, 2014), then such grassroot media projects, in particular those that make use of and form in online spaces, are able to meet the information needs of this community of concern, where access to specific localised news and content may not be available in their preferred language or deal with relevant issues or concerns. Risk reduction and recovery, as is well established in the literature, often fail to attract the attention from mainstream news media that is afforded to the acute disaster impacts. By providing access to news and information that are not available through other media, such grassroots and citizen-initiated initiatives are able to help sustain this community of concern but may also contribute to forms of participation. This is valuable in disaster contexts since it can encourage both formal and informal engagement from these broader communities of concern with disaster recovery. This will take different forms but could include offering support for or comment on reconstruction initiatives, engaging in forms of advocacy such as sharing news or stories that raise awareness of underreported issues to establishing links and reconnecting with recovering communities, or by directly contributing to recovery efforts, for example by volunteering, investing in or visiting the area. Ultimately, such engagement may be valuable due to its potential to contribute to recovery processes and empower communities to make adaptations to reduce disaster risks and address vulnerabilities.

References

Adelstein, J., & Stucky, N. (2013). Japan's Ishinomaki city still stricken from Tsunami. Retrieved April 1, 2019, from https://www.thedailybeast.com/japans-ishinomaki-city-still-stricken-from-tsunami.

Cottle, S. (2014). Rethinking media and disasters in a global age: What's changed and why it matters. *Media, War & Conflict, 7*(1), 3–22. https://doi.org/10.1177/1750635213513229

Harlan, C. (2012). A year after Japan's triple disaster, an uncertain recovery. *The Washington Post.* Retrieved April 1, 2019, from https://www.washingtonpost.com/world/asia_pacific/a-year-after-japans-triple-disaster-an-uncertain-recovery/2012/03/05/gIQAcR920R_story.html?noredirect=on&utm_term=.5384fd711f37.

Houston, J. B., Pfefferbaum, B., & Rosenholtz, C. E. (2012). Disaster news framing and frame changing in coverage of major US natural disasters, 2000–2010. *Journalism & Mass Communication Quarterly, 89*(4), 606–623. https://doi.org/10.1177/1077699012456022

Japan Fire and Disaster Management Agency. (2016). Retrieved October 1, 2018, from https://www.fdma.go.jp.

Kurpius, D. D., Metzgar, E. T., & Rowley, K. M. (2010). Sustaining hyperlocal media. *Journalism Studies, 11*(3), 359–376. https://doi.org/10.1080/14616700903429787

Mannakkara, S., Wilkinson, S., Willie, M., & Heather, R. (2018). Building back better in the Cook Islands: A focus on the tourism sector. *Procedia Engineering, 212*, 824–831. https://doi.org/10.1016/j.proeng.2018.01.106

Matthews, J. (2019). Cultural otherness and disaster news: The influence of western discourses on Japan in US and UK news coverage of the 2011 Great East Japan Disaster. *International Communication Gazette, 81*(4), 372–392. https://doi.org/10.1177/1748048518774982

McMorran, C. (2017). From volunteers to voluntours: Shifting priorities in post-disaster Japan. *Japan Forum, 29*(4), 558–582. https://doi.org/10.1080/09555803.2017.1307257

NHK World. (2013, October 23). *Newsline*. Retrieved October 1, 2018, from https://www.youtube.com/watch?v=H0Qa-d45eNk.

Pantti, M., Wahl-Jorgensen, K., & Cottle, S. (2012). *Disasters and the media*. New York: Peter Lang.

Ploughman, P. (1995). The American print news media 'construction' of five natural disasters. *Disasters, 19*(4), 308–326.

Reader, B., & Hatcher, J. A. (2012). *Foundations of community journalism*. Thousand Oaks: SAGE Publications.

Robinson, S. (2014). Introduction, community journalism midst media revolution. *Journalism Practice, 8*(3), 113–120. https://doi.org/10.1080/17512786.2013.859822

CHAPTER 5

Drought Is a Disaster in the City: Local News Media's Role in Communicating Disasters in Australia

Jacqui Ewart

Australia frequently experiences severe natural disasters including bush fires, droughts, storms and floods and sometimes all four simultaneously in different parts of the country. As the finishing touches were being made to this chapter in January-February 2019, Australia was facing a series of natural disasters: fires were burning in the state of Tasmania, parts of Queensland were experiencing severe flooding, while an ongoing drought-affected other areas of the country. Images of drought-starved cattle that were drowning as a result of monsoonal weather in Queensland flew around the country. In mid-2018, the drought had come to the attention of those living in Australia's cities when *The Guardian*'s Australia edition ran a series of stories about the effects of the drought in parts of New South Wales and Queensland. News reports quickly followed that farmers in drought-affected areas of New South Wales were sending livestock to abattoirs because they could not afford to feed them (Cox, 2018; Morphett,

J. Ewart (✉)
Griffith University, Brisbane, QLD, Australia
e-mail: J.Ewart@griffith.edu.au

J. Matthews, E. Thorsen (eds.), *Media, Journalism and Disaster Communities*, https://doi.org/10.1007/978-3-030-33712-4_5

2018). The impact of the drought was brought home to readers of *The Telegraph*, a Sydney-based tabloid newspaper, when it published a story about a farmer, who was preparing to shoot 1200 starving sheep. That story was accompanied by images showing a dry, grassless paddock with wasted sheep and an image of Les the farmer with an emaciated lamb. National and state-based news media quickly took up the story of Les and other farmers' plights. The significant attention focused by state and national media on the drought and its impact on farmers made it seem like it was a recent crisis despite the fact that it had unfolded over the past decade or more. The extensive coverage afforded by state and national news media of the drought highlighted an important gap in the research, namely that we know relatively little about how Australia's local news reported the drought.

Given the complex interplay of natural hazards in Australia, it is not surprising that disaster communication in the Australian context has attracted significant research attention. Despite the addition of social media to the suite of disaster communication tools, traditional news outlets and media continue to be important tools for information dissemination. This chapter explores the distinct ways in which local, state and national news media in Australia approach the coverage of natural disasters.

Those affected by disasters, both as they unfold and in their aftermath, rely on a range of news and information sources based within or focused on their geographical location. Research on the use of local news sources including local radio, newspapers and community media has revealed the crucial and often underappreciated role such media fulfil at times of natural disasters. These information sources are considered highly reliable by their audiences, providing the type of news and information that they most need, and are also able to support disaster planning and response.

There are clear differences in the approaches taken by local, state and national news media to covering disaster events, with the research highlighting the crucial and often under-appreciated role that local news media in Australia play in keeping their communities informed before, during and after natural disasters.

The Australian News Media Context

There are two things that set Australia apart from many other Western countries in relation to news media coverage of natural disasters. The first is that the national, publicly funded broadcaster ABC, which provides a range of national, state and local news services, works closely with emergency services to 'deliver warnings, alerts, information and news about disasters and emergencies on TV, Radio, online and on mobile' (Radioinfo, 2006, n.p.). Delivery of these services is enshrined in the Broadcasting Services Act 1992. The second is that community broadcasting also has an obligation under the Broadcasting Services Act (1992) to provide emergency warnings in the lead up to and during disasters. Local, state and national news media have very different but important roles in communicating about disasters, and in order to understand these a brief overview of Australia's media terrain is provided.

Australia's news media environment is characterised by a mix of commercial and non-commercial media located in metropolitan, regional and rural areas. Some are government funded, others are supported by advertising, subscriptions or a mix of both. Fairfax, one of Australia's largest news media organisations reported in 2018 that much of its revenue came from news and advertising, events, digital marketplaces such as Domain and a company called Stuff in New Zealand. News Corp has similar event and diversified arrangements such as streaming FoxTel (Fairfax, 2018a). Just over 72% of Australia's daily newspapers based in capital cities are owned by News Corp Australia, wholly owned by US-based News Corporation (Holmes & Star, 2018). The next most powerful and pervasive organisation is Nine. A $4 billion merger in 2018 of Nine Entertainment and Fairfax Media was made possible by a suite of new media laws passed by the Australian Federal Parliament in 2017 (Australian Government, 2017) and was implemented on December 7, 2018 (Fairfax, 2018b). The new Nine organisation provides television, radio, newspaper and online news content as well as streaming (Stan) and documentary products.

Significantly, as well as their multimillion capital city and metropolitan publications, both News Australia and the former Fairfax Media (now known as Nine Publishing) own and control stables of regional and local community newspapers, radio and television stations and websites that reach into the kitchens, loungerooms and verandas of millions more Australians, especially those in areas at risk from natural hazards, such as bush fire, cyclones and drought.

In Western Australia, the Community News subsidiary operates 17 local (Community News, n.d.) newspapers around Perth, in South Australia the Messenger subsidiary operates 7 (News Corp Australia, 2018a) around Adelaide, Quest operates 13 community newspapers in south-east Queensland, the Leader group operates 26 community newspapers around Victoria (News Corp Australia, 2018b) (including near-metro Melbourne) and the New South Wales 'NewsLocal' brand accounts for 19 mastheads in suburban Sydney and regional parts of the states. Outside the capitals, News Australia publishes 61 regional and small-town newspapers in Queensland, 11 in New South Wales, 1 in Tasmania and South Australia, and 2 in the Northern Territory (News Corp Australia, 2018c).

For its part, Nine Publishing (the former Fairfax arm of Australian Community Media and Printing) includes 'more than 160 regional publications and community-based websites' and 'approximately 130 community-based websites' (Nine Digital Pty Ltd., 2018).

Outside the so-called mainstream news outlets there are many more privately owned community or local news providers in Australia. A study of all Australian newsrooms operating in 2013 (Cokley, Gilbert, Jovic, & Hanrick, 2016, pp. 63–64) reported that for a total of 11,635 journalists employed across the range of print, broadcast, online and agencies in Australia, the majority (6104) were employed in organisations that employed nine or fewer journalists.

A mostly not-for-profit community broadcasting sector is comprised of radio stations, some television and online news. Most community radio stations provide information to geographically specific locations or to communities of like-minded people, performing an important role in providing information and news about natural disasters. In 2017, media reforms passed by Australia's Federal Parliament were aimed at strengthening and diversifying 'our media industry and to support local journalism jobs in regional areas across the country' (Australian Government, 2017). The reforms also included the implementation of 'a $60 million Regional and Small Publishers Jobs and Innovation package' (Australian Government, 2017, n.p.) designed to provide financial grants to publishers of 'a newspaper, magazine or other periodical, or to a content service provider' located in regional areas. The grants aimed to support growth and innovation in civic and public journalism (Commonwealth of Australia, 2018) and included scholarships for regional journalism students.

National and State-based News Media and Australian Disasters

Historically, research into Australian natural disasters have tended to focus on mainstream news media coverage of disasters such as fires, floods, storms and drought (Cohen, Hughes, & White, 2006; Ewart, 2002;

Gawenda & Muller, 2009; Hughes & White, 2004; Leitch & Bohensky, 2014; McDougall, 2011; Ward, 2005) and the information sources audiences accessed when disasters occur (Cretikos et al., 2008). However, very few studies have focused on the role of local news or community media in information provision during the various phases of natural disasters.

An early study by McKay (1983) explored the themes in three Adelaide newspapers' coverage of the Ash Wednesday bush fires, which occurred in the Australian states of South Australia and Victoria in 1983. She identified seven common themes in their reportage, including: warning and response information; personal hardship and heroism stories; fund-raising schemes; general descriptions of loss; the search for who or what was to blame; land use and planning issues; and how funds were raised and government relief schemes were allocated. McKay (1983) analysed the amount of space given to each of these themes by those newspapers revealing that in the lead up to the fires they devoted very little coverage to warnings or to how residents should prepare for and respond to the extreme fire danger. In the wake of the fires, there was a period of significant interest and attendant coverage in these newspapers but that interest rapidly declined. McKay identified that the newspapers lost interest in the story quickly and their attention shifted to the search for blame. She concluded that there was a lost opportunity to provide communities affected by the fires with information that would mitigate similar events in the future.

A study by Hughes and White (2004) reviewed research into reportage of bush fires in Australian news media and the official inquiry reports that followed. They found that some Australian news media focused on the search to attribute blame for the fires. Hughes and White revealed that the research and inquiry reports showed that local radio was sometimes the only source of information for disaster communities. Ewart and McLean (2015) also examined how the narrative of blame played out in a state-based newspaper and a national newspaper. They undertook a thematic analysis of the reportage by two major newspapers of the release of the findings of a Commission of Inquiry into the severe floods that occurred in South-East Queensland in late 2010 and early 2011. The key themes in that coverage were those of failure and the search to apply responsibility, despite the absence of blame in both of the Commission's reports. The theme of blame emerged primarily in opinion and editorials, while that of failure was largely confined to news and feature articles. However, Ewart and McLean (2015) concluded that the search to lay blame thwarted a discussion about how a similar disaster might be prevented in the future. More recent research by Anderson, Chubb, and Djerf-Pierre (2018) examined how the blame game emerged in national, state and local news coverage of the Black Saturday bush fires in 2009 in Australia. In a framing analysis of state, national and local news media coverage of the fires they found that the focus on appropriating blame obscured the deeper complexities of bush fires and the nexus between media coverage and policy responses. Their findings reveal differences in how local newspapers and some national media pursued the blame game. For example, there was a more balanced approach to covering environmentalists and environmental issues in the local newspapers in their study as well as in a left-leaning national newspaper, compared to a more conservative, right-leaning national newspaper. There was a lack of reporting that explored the deeper causal factors associated with the fires across all newspapers they studied.

Another important theme in the research is the problematic nature of journalistic practices at sites of natural disasters. For example, Bilboe (1998) examined the practices of journalists reporting on a major landslide in the ski resort of Thredbo in the Australian state of New South Wales. She focused on the ethical breaches that occurred amongst some of the journalists covering the story, identifying that reporters and camera crew working for non-local news media were more likely than local journalists to take risks, breach and invade the privacy of those affected and those living in the community involved. Her findings will be further explored later in this chapter. Similar themes emerged in relation to the practices of journalists from state-based news media in a study into the practices of journalists covering the aftermath of bush fires at the small Australian seaside community of Tulka in South Australia (Ewart, 2002). Those reporting the fires were not local journalists but were from television stations and newspapers based in Adelaide, the capital of the state of South Australia. That study identified some of the problematic practices journalists engaged in when covering the aftermath of that fire. Cohen, Hughes, and White (2007) also identified key differences in how local and non-local journalists approached reporting natural disasters.

The research that has focused on national and state-based media, highlights how reporting from these outlets tends to focus on the dramatic aspects of disaster impacts, with stories accompanied by shocking images and coverage paying little attention to disaster recovery. In contrast, local news are able to provide vital information for disaster communities, with reporters being more attuned to their own communities than their state and national counterparts. Local news coverage is also able to help those affected by disaster, with audiences reporting high levels of trust in local media as a consequence. These differences will be elaborated on later in this chapter.

Drought is another under-researched area of news media coverage of natural disasters. The few studies that have explored this issue have identified that unless there are specific events with associated images, coverage is sporadic at best. This can be explained by the slowly unfolding nature of prolonged drought in Australia not fitting the immediacy needs of many news organisations. For example, Bell (2009) examined two newspapers' coverage of a drought in Sydney, Australia, in 2002. Her focus was on how Sydney newspapers, which provide coverage of the state rather than specific localities, represented the drought and its effects on Sydney, rather than on farming land in New South Wales. Another important study by Ward (2005, p. 93), which examined two national newspapers' coverage of drought between 1997 and 2002, revealed that coverage was irregular. He argued that coverage of drought is largely event driven because of the need for images and as a complex hazard develops slowly over time, which runs counter to the news value of immediacy.

There is a relationship between news media use and how connected people feel to their community. In his study of an Australian regional newspaper, Killiby (1994) highlighted that those community members who use regional and country newspapers have stronger ties to their communities than those who do not, identifying how the use of local news media strengthened the sense of connection for readers to their communities. Stamm (1985) highlighted the important contribution local news media made to community formation, and Pretty (1993, p. 112) found that journalists on non-country dailies had a 'greater knowledge of and higher regard for their readers than other newspaper journalists.'

Local News Media and Australian Disasters

There is relatively little research into how local news media in Australia cover disaster, but there are tantalising hints in the small body of research that local news and information sources have a valuable role for their audiences. Some differences have emerged in the findings of research about local media's role in both natural and anthropogenic disasters. For example Oh, Agrawal, and Rao (2013) suggest that following a disaster people turn to trusted institutional sources of information such as mainstream news media, but when they are unable to access relevant information, they turn to local news outlets as well as other sources (Oh et al., 2013). Other researchers, however, have identified that disaster-affected communities trust local sources of news and information and will access these first in the event of a natural disaster (Ewart & Dekker, 2013; Perez-Lugo, 2004). Anthony and Sellnow (2011), for example, recognised that local media sources provided crucial information about protection during and recovery following Hurricane Katrina. While some of the aforementioned research has focused on countries other than Australia, it points to similarities in the importance of local news media and local media more broadly for those affected by natural disasters.

Before turning to how local news cover natural disasters, it is important to acknowledge that the definition of local news and information is somewhat contested. Hess and Waller (2014, p. 121) noted the importance of the 'sense of place' that local news media provide for their audiences but argued that local news media outputs, also referred to as community news media, should now be termed 'geosocial journalism' because of the changes associated with the digital news environment. They also suggested that researchers should reconsider their use of community as the 'dominant frame of reference for describing and discussing commercial newspapers that serve small towns and cities across the western world' and, instead, that the term 'geosocial journalism' may be more appropriate as it facilitates researchers' understanding of the place that geosocial publications hold and the roles they fulfil.

In another study of three newspapers located in small cities and towns in regional Australia, Hess and Waller (2016) found that local news involved a deep and continuing connection with, and location within, a community. They (2016, p. 264) explained that 'to be local is to have a grounded connection with, and understanding of, a physical place and its social and cultural dimensions that is practical and embodied.' The role of local news, Hess and Waller (2016, p. 264) argued, 'involves an investment of time, requiring that one maintains a prolonged and continual presence in that place.'

Researchers have considered how local and community news media are used by audiences after disaster. To that end, Cohen et al. (2006) evaluated how residents of the Grampians region in Australia used different forms of media during bush fires that affected the area in 2006 and highlighted the importance of local knowledge for journalists when providing warnings about these types of disasters and that rural and regional news sources were about helping their audiences prepare and survive such disasters. Cohen and colleagues also suggested that their findings would help emergency response agencies understand how residents access and use media sources with the potential of improving engagement between agencies and media sources. Significantly, participants in their study (2006, p. 4) described local regional media, as 'more knowledgeable about fire preparedness, fire mitigation and the importance of delivering specific warnings during incidents. In contrast, urban and national media, despite attempts by fire agencies to work with and 'manage the media' in various other ways, will tend to depict large bushfires as 'disasters.' National media, and in particular commercial television, they argued, will tend to sensationalise bushfires in promoting their own agendas and their own commercial interests.'

The concept of hyperlocal news is also useful when considering how audiences contribute news and information about these events. Picone (2007, p. 102) defined hyperlocal news as 'devoted to the stories and minutiae of a particular neighbourhood, ZIP code or interest group within a certain geographic area.' Hyperlocal news may be important during disasters because it offers information that would not be considered newsworthy in the traditional sense, but can be of great value for communities. Two studies illustrate this point. The first concerned non-populist talkback radio programmes in Australia, identifying that the aforementioned programmes provided local news and information that other news outlets did not necessarily recognise as news (Ewart & Dekker, 2013). They pointed to the importance of talkback radio in relation to the provision of information about natural disasters including major floods, storms and other emerging weather events. Ewart and Dekker (2013) noted that media have different agendas when covering these types of disasters and that public service media tended to focus on providing immediate information for disaster communities. This created a high level of trust between audiences and the media they were using and reflected earlier findings by Perez-Lugo (2004), who identified that local media are able to foster a sense of connection for their audiences and provide emotional support and vital information when disasters strike.

While talkback radio in Australia has no official mandate to provide emergency warnings and information, with the Australian Broadcasting Service as the official provider of emergency information, the importance of local Australian Broadcasting Corporation (ABC) radio was exemplified by two studies. The first was undertaken by Cretikos et al. (2008, p. 1) who found that the local ABC radio played 'a key role in disseminating public health advice' during severe storms that hit the Hunter Valley in the Australian state of New South Wales in June 2007. The second, by North and Dearman (2010, p. 80), explored the Gippsland ABC radio station's use of audience-generated content during the Black Saturday bush fires in Victoria in 2009. Through interviews with several staff who worked at the radio station, North and Dearman (2010, p. 83) revealed that the ABC's role as emergency broadcaster via the services it provided during disasters through local radio was closely connected to its attempts to 'redefine itself as having a strong community focus.' During the fires the station opened the phone lines to callers to share their stories and this empowered community members as they became newsmakers. Importantly, this case study exposed the shifting terrain of local news production and how journalists and audiences were engaging with radio.

The critical role of local news and information sources during natural disasters was also touched on in one of the first audience studies of community broadcasting in Australia. Meadows, Forde, Ewart, and Foxwell (2007) identified two examples of community radio stations that were crucial information sources for regional and rural communities during natural disasters. The first was in the town of Katherine in the Northern Territory when the local community radio station continued to broadcast in dire circumstances during a major flood in 1998. Meadows et al. (2007, p. 36) explained that a 'lack of local media, which focused on local issues, meant there was a high level of reliance on the community radio station for practical information such as how quickly the river was rising.' The radio station also connected local residents with each so that they were able to provide equipment and assistance to those in need. The second example they described was in Tumut in New South Wales where bush fires regularly occur. The participants in Meadows and colleagues' study (2007, p. 36) identified that Tumut's community radio station 'provided the sort of information that commercial stations are not interested in' when fires occurred.

The aforementioned research highlights the importance of local radio in the provision of news and information to those communities affected by natural disasters. Local newspapers, however, also have a role in covering these types of disasters. Leitch and Bohensky (2014), for example, examined how the discourse of resilience was represented through state and local newspapers' coverage of natural disasters. Their study identified that journalists from state and national media used the term 'resilience' more often than journalists from a local paper when referring to their own, affected community. Another important finding from that study was about the sources used in disaster reporting. Leitch and Bohensky (2014, pp. 18–19) revealed that those sources 'who referred to the resilience of a specific community were generally from beyond the local affected community: for instance, the politicians or government sources were generally from a state or national scale rather than members of an affected community.' The research has shown that local newspapers are much more likely to include local voices (Ewart & Massey, 2005) and are more attuned to their communities.

Research into the practices of journalists during natural disasters has revealed differences in how local and non-local journalists cover these events. Bilboe (1998) summed up these differences in her study of ethical breaches by journalists reporting a major landslip at Thredbo, a ski resort in south-eastern Australia in 1997. She wrote (1998, p. 91) that ideas about 'community are also important in covering disasters. Often local news media are more aligned with community perceptions and concerns compared to national journalists from other locations or countries sent to cover the disaster. Clear differences in reporting practices are evident with local media generally more sensitive to their own communities.' Bilboe provided a highly illustrative example of how local journalists' sensitivity to their community resulted in very different outcomes than those experienced by state, national and international journalists. She (1998, p. 98) explained: 'Within a day, some members of the Thredbo community found the journalists an unwelcome intrusion. Signs appeared in the village, particularly at the petrol station, that members of the media would not be served, except those from the local radio station. Journalists were ignored in village shops and many ended up staying further away from the village (towards the slip site) and sending others for supplies.' Bilboe's findings correspond with the results of Pretty's (1993) study, which highlighted that journalists working for country and regional newspapers had a closer relationship to their readers than their metropolitan counterparts did to theirs. More recently, Muller (2010) examined the often-traumatic experiences of journalists covering bush fires that occurred in Australia in 2009. However, he did not differentiate between the experiences of national, state and local journalists. A more recent study by Ewart and Mclean (2018) sought to go beyond identifying poor practice to offer a series of recommendations for journalists when covering disasters. They distilled the international literature and key points emerging from interviews with senior emergency managers from around the world to present a list of best practice approaches for journalists reporting on disasters. They noted (2018, p. 3) that 'local, national and international media will emphasise different aspects of the best practice checklist when reporting disasters because there are distinctions in the roles of local journalists and those journalists observing and reporting on the disaster from a distance.'

Discussion and Conclusion

This chapter has explored some of the key studies that have been undertaken into news media coverage of natural disasters in Australia. In examining that literature, this chapter has argued that researchers have primarily focused on national and state news media coverage of natural disasters in Australia, largely to the detriment of studying local news media coverage of these events. The small amount of research that has been undertaken into local news and information sources around natural disasters reveals the important role these media fulfil for affected communities. It is clear that local media outlets often provide information that is not considered newsworthy by state and national news media. That information provides an important, if not under-recognised, lifeline for people involved in natural disasters. Distinctive types of news media and media more generally, for example talkback radio and community broadcasting, perform crucial roles for those caught in natural disasters. In addition, the research evaluated through this chapter indicates that those involved in disaster planning and response in Australia may be able realise the untapped potential of local news media to provide information about hazards and disaster risks and to those affected by natural disaster. For disaster management agencies, here specifically those under the umbrella of Emergency Management Australia, the ability to access local news media is crucial in communicating about disaster risk, response and recovery. There are also considerations for emergency management agencies in relation to understanding the importance of local media outlets to disaster communities. Disaster communities seek information from talkback radio and community broadcasting playing important roles, and it is therefore important to enhance the connections between disaster management agencies and local media. This could, for example, mean involving local news media staff in disaster planning exercises or providing a seat at a disaster control centre's command table for those who provide local news, which may create new possibilities for cooperation and for disaster communities (Ewart & McLean, 2018).

Emergency management organisations might also consider expanding their engagement with these spaces.

This chapter raises some important issues for researchers, policymakers, emergency management agencies and media organisations. Researchers need to look beyond the national and state level to examine the role local news media and information sources perform for those who desperately need localised information when natural disasters occur. More recently, research has shifted focus to consider the use of social media in the various phases of disasters, but that may be at the expense of research into the role of local media and locally produced news. Further research would be useful as to compare the news media sources used by those from with disaster-affected communities and those who are distant from events. A number of considerations and implications for policy emerge from the critical review of the literature presented in this chapter. These are in part related to the 2017 media reforms introduced by the Australian government. At the time of writing it is not yet clear how those reforms will fully play out but ensuring that local news media, especially in regional and rural areas, survive and thrive in a restricted media environment is crucial as in Australia it is these areas that are most at risk from natural hazards. Finally, for media organisations the literature highlights that no one form of news media or media more generally can provide all audiences with everything they need when natural disasters occur. This only emphasises the need for a range of media in an environment that experiences severe and frequent natural disasters, whether they unfold slowly or are sudden-impact events.

In Australia, where communities can face multiple natural disasters in a relatively short period of time, access to timely and locally relevant information for those at risk from and living through disaster is vital.

References

Anderson, D., Chubb, P., & Djerf-Pierre, M. (2018). Fanning the blame: Media accountability, climate and crisis on the Australian "Fire Continent". *Environmental Communication*. https://doi.org/10.1080/17524032.2018.1424008

Anthony, K. E., & Sellnow, T. L. (2011). Information acquisition, perception, preference, and convergence by Gulf Coast residents in the aftermath of the Hurricane Katrina crisis. *Argumentation and Advocacy, 48*, 81–96.

Australian Government. (2017). Australian media reforms passed by Parliament: The Government has passed historic changes to Australia's media laws—The largest reforms in nearly three decades. Department of the Arts and Communication. Retrieved October 1, 2018, from https://www.communications.gov.au/departmental-news/australian-media-reforms-passed-parliament.

Bell, S. (2009). The driest continent and the greediest water company: Newspaper reporting of drought in Sydney and London. *International Journal of Environmental Studies, 66*(5), 581–589.

Bilboe, W. (1998). The Thredbo landslide: Was it only media ethics that came tumbling down. *Australian Journalism Review, 20*(2), 88–110.

Cohen, E., Hughes, P., & White, P. (2006). Bushfires and the media: Report No 4: Media and bushfires: A community perspective Grampians fires 2006. Melbourne La Trobe University. Retrieved October 1, 2018, from https://www.latrobe.edu.au/__data/assets/pdf_file/0006/215979/report-4.pdf.

Cohen, E., Hughes, P., & White, P. (2007). Media and bushfires: A community perspective of the media during the Grampians fires 2006. *Environmental Hazards, 7*(2), 88–96.

Cokley, J., Gilbert, L., Jovic, L., & Hanrick, P. (2016). Growth of 'Long Tail' in Australian journalism supports new engaging approach to audiences. *Continuum, 30*(1), 58–74.

Commonwealth of Australia. (2018). Environment and Communications Legislation Committee, Communications Legislation Amendment (Regional and Small Publishers Innovation Fund) Bill 2017, Canberra.

Community News. (n.d.). About us. Retrieved February 18, 2019, from https://www.communitynews.com.au/about-us/.

Cox, L. (2018, June 22). Australia doesn't realise': Worsening drought pushes farmers to the brink. *The Guardian*. Retrieved from https://www.theguardian.com/environment/2018/jun/10/australia-doesnt-realise-worsening-drought-pushes-farmers-to-the-brink

Cretikos, M., Eastwood, K., Dalton, C., Merritt, T., Tuyl, F., Winn, L., & Durrheim, D. (2008). Household disaster preparedness and information sources: Rapid cluster survey after a storm in New South Wales, Australia. *BMC Public Health, 195*(8), 1–9.

Ewart, J. (2002). Prudence not prurience: A framework for journalists reporting disasters. Retrieved June 5, 2003, from https://www.researchgate.net/publication/228793152_Prudence_not_prurience_A_framework_for_journalists_reporting_disasters.

Ewart, J., & Dekker, S. (2013). 'Radio, someone still loves you!' Talkback radio and community emergence during disasters. *Continuum: Journal of Media and Cultural Studies, 27*(3), 365–381.

Ewart, J., & Massey, B. (2005). 'Local people mean the world to us': Australia's regional newspapers and the 'closer to readers assumption'. *Media International Australia Incorporating Culture and Policy, 115*(1), 94–108.

Ewart, J., & McLean, H. (2015). Ducking for cover in the 'blame game': News framing of the findings of two reports into the 2010–11 Queensland floods. *Disasters, 39*(1), 166–184.

Ewart, J., & Mclean, H. (2018). Best practice approaches for reporting disasters. *Journalism*. https://doi.org/10.1177/1464884918757130

Fairfax Media. (2018a). Fairfax Media annual report 2018: Going for growth. *Fairfax Media*. Sydney, Australia. Retrieved October 1, 2018, from https://www.fairfaxmedia.com.au/ArticleDocuments/193/2018%20FXJ%20Annual%20Report_FINAL_14AUG2018.pdf.aspx?Embed=Y.

Fairfax Media. (2018b). Fairfax/Nine merger. Retrieved February 9, 2019, from https://www.fairfaxmedia.com.au/Investors/Welcome/Fairfax/Nine-Merger/nine-merger?Embed=Y.

Gawenda, M., & Muller, D. (2009). *The Black Saturday bushfires: How the media covered Australia's worst peace-time disaster*. Melbourne: Centre for Advanced Journalism, University of Melbourne.

Hess, K., & Waller, L. (2014). Geo-Social Journalism: Reorienting the study of small commercial newspapers in a digital environment. *Journalism Practice, 8*(2), 121–136.

Hess, K., & Waller, L. (2016). River flows and profit flows: The powerful logic driving local news. *Journalism Studies, 17*(3), 263–276.

Holmes, D., & Star, C. (2018). Climate change communication in Australia: The politics, mainstream media and fossil fuel industry nexus. In: W. Leal Filho, E. Manolas, A. Azul, U. Azeiteiro, & H. McGhie (Eds.), *Handbook of climate change communication: Vol. 1. Climate change management* (pp. 141–170). Cham, Switzerland: Springer.

Hughes, W. P., & White, P. B. (2004). Bushfires and the media: A preliminary literature review. Retrieved October 1, 2018, from http://www.bushfirecrc.com/sites/default/files/managed/resource/report-one.pdf.

Killiby, C. (1994). Civic attitudes and regional media use. *Australian Journalism Review*, in Journalism, *16*(1), 81–87.

Leitch, A. M., & Bohensky, E. L. (2014). Return to 'a news normal': Discourses of resilience to natural disasters in Australian newspapers 2006–2010. *Global Environmental Change, 26*, 14–26.

McDougall, K. (2011). Understanding the impact of volunteered geographic information during the Queensland floods. Retrieved October 1, 2018, from https://www.researchgate.net/profile/Li_He11/publication/260165261_On_the_Virtual_Moon_Framework_and_the_Virtualization_Technology_of_Chang'e-1_Lunar_Satellite_Data/links/0deec52fd793934bbd000000.pdf#page=84.

McKay, J. M. (1983). Newspaper reporting of bush-fire disaster in south-eastern Australia—Ash Wednesday 1983. *Disasters, 7*(4), 283–290.

Meadows, M., Forde, S., Ewart, J., & Foxwell, K. (2007). *Community media matters: An audience study of the Australian community broadcasting sector*. Brisbane: Griffith University.

Morphett, J. (2018, July 22). Silence of the lambs: NSW farmer to shoot starving flock because he can't afford to feed them. *The Sunday Telegraph*. Retrieved from https://www.dailytelegraph.com.au/technology/science/silence-of-...-afford-to-feed-them/news-story/73cb7f3e90525214ae89cab3b613d856.

Muller, D. (2010). Ethics and trauma: Lessons from media coverage of Black Saturday. *The Australian Journal of Rural Health, 18*, 5–10.

News Corp Australia. (2018a). Messenger. Retrieved February 18, 2019, from https://www.newscorpaustralia.com/brand/messenger-news/.

News Corp Australia. (2018b). Leader. Retrieved February 18, 2019, from https://www.newscorpaustralia.com/brand/leader-community-news/.

News Corp Australia. (2018c). Regional media. Retrieved February 19, 2019, from https://www.newscorpaustralia.com/networks/news-nrm/.

Nine Digital. (2018). Australian community media & printing. Retrieved February 18, 2019, from https://www.nineentertainmentco.com.au/brand-australian-community-media-and-printing.

North, L., & Dearman, P. (2010). The rhetoric of 'community': ABC local radio's coverage of the 2009 Victorian bushfires. *Media International Australia, 137*, 80–89.

Oh, O., Agrawal, M., & Rao, H. R. (2013). Community intelligence and social media services: A rumour theoretic analysis of tweets during social crises. *Management Information Systems Quarterly, 37*(2), 407–426.

Perez-Lugo, M. (2004). Media uses in disaster situations: A New focus on the impact phase. *Sociological Inquiry, 74*(2), 210–225.

Picone, I. (2007). Conceptualising online news use. *Observatorio Journal, 3*, 93–114.

Pretty, K. (1993). Dusting off the grassroots: A survey of Australian country journalists. *Australian Studies in Journalism, 2*, 83–87.

Radioinfo. (2006). ABC774 Manager Ian Mannix moves to new emergency services national role. Retrieved October 2, 2018, from https://radioinfo.com.au/news/abc774-manager-ian-mannix-moves-new-emergency-services-national-role.

Stamm, K. R. (1985). *Newspaper use and community ties: Toward a dynamic theory*. Norwood, NJ: Ablex Publishing Corporation.

Ward, I. (2005). Drought, news media and policy debate. In L. C. Botterill & D. A. Wilhite (Eds.), *From disaster response to risk management: Australia's national drought policy* (pp. 85–97). Dordrecht: Springer.

CHAPTER 6

Media and Climate Migration: Transnational and Local Reporting on Vulnerable Island Communities

Anna Roosvall, Matthew Tegelberg, and Florencia Enghel

Climate change and migration are two core challenges of our times. Increasingly, they merge and constitute climate-induced migration, which is a global challenge that affects local communities unevenly. Thus, it must be discussed as an issue of *climate justice* (Dreher & Voyer, 2015). Studying *media reporting* on climate-induced migration is also crucial since people often learn about this accelerating, inequitably distributed phenomenon via journalism. We focus on media reporting regarding three US islands, all hard-hit by climate change and related extreme weather, exposed to recent or impending climate migration: Sarichef Island, Alaska, home to Iñupiaq indigenous people; Isle de Jean Charles, Louisiana, home to

A. Roosvall (✉)
Stockholm University, Stockholm, Sweden
e-mail: anna.roosvall@ims.su.se

M. Tegelberg
York University, Toronto, Canada

F. Enghel
Malmö University, Malmo, Sweden

J. Matthews, E. Thorsen (eds.), *Media, Journalism and Disaster Communities*, https://doi.org/10.1007/978-3-030-33712-4_6

Biloxi-Chitimacha-Choctaw Native American people; and Puerto Rico. Similarities and differences between these three cases will help us explore and understand variations in the reporting.

There are high degrees of poverty on each of these islands. In Sarichef Island and Isle de Jean Charles, indigenous peoples have specific indigenous status to varying extents and tend to lack proper political representation (Roosvall & Tegelberg, 2018). Puerto Ricans have a politically weak status and lack proper political representation inasmuch as the island is controlled by the US Congress (see the section *Puerto Rico*). These issues correspond to Fraser's (2008) three dimensions of injustice: maldistribution (of economic means), misrecognition (of identity/status) and misrepresentation (lack of political representation). In terms of community, the selected cases furthermore correspond to geographical scales of justice in varying ways (Fraser, 2008). The indigenous communities on Sarichef Island and Isle de Jean Charles are *local communities* that are ethnically homogenous, while Puerto Ricans belong to a *regional community* that diverges in terms of class and ethnicity. The three communities relate to the same *nation*—the US—and are affected by *global* climate change, but in diverse ways. Isle de Jean Charles is drowning, Sarichef Island is hard-hit by erosion and melting ices, and Puerto Rico has been hit by several climate change enhanced hurricanes, most famously María in 2017. We are interested in similarities and differences connected to these varying geographies, in their differing political and cultural contexts, and in how reporting on them varies in different media contexts: local versus transnational media.

The aim of this study is thus to explore how (imminent) climate migration from three US islands is understood in local and transnational reporting, and particularly how issues of justice inform reporting on places, peoples and perceived problems in the examined articles.

RQs:

1. How is the connection between climate change and migration understood in the coverage and what does it have to do with media geography (local/transnational news), political geography (status issues, representation issues) and natural geography (island vulnerability, islands as parts of a global problem, etc.)?
2. How are migrants represented concerning redistribution, recognition and (political) representation (Fraser's three dimensions of injustice) in local versus transnational journalism?

Media, Climate Migration and Climate Justice

Media coverage of climate-induced migration is severely under-researched. Dreher and Voyer (2015) studied how climate migration in Small Island Developing States (SIDS) was covered in Australian media and how Pacific Islanders relate to this coverage, applying a climate justice lens. They found that justice aspects were rare in the four main media framings identified: (1) as evidence of climate change; (2) as climate change victims; (3) as climate refugees; and (4) as travel destinations. Interviewees proposed alternative frames that would align better to their situations, needs and perspectives and allow climate justice to be evoked more pertinently: for example, human rights and migration with dignity frames. Other studies refer to climate migration and various forms of communication more generally, but not to journalism (e.g. Farbotko & Lazrus, 2012).

In an overview of the climate–migration nexus, Piguet, Pécoud, and de Guchteneire (2011, p. 1) write that 'the extent to which the environment determines migration is intimately connected to the status to be associated with the people concerned'. A crucial problem is that those affected are not generally recognized as constituting *the status of refugees* (Betts, 2013). Depending on who is forced to migrate, there are also diverging notions on whether migration is a sign of failed adaptation to climate change or a form of creative adaptation (Bardsley & Hugo, 2010). There are for instance racial underpinnings of climate-induced migration (Baldwin, 2016). These are relevant, to varying degrees, in relation to the internal domestic displacement of indigenous groups and people fleeing Puerto Rico for the US mainland.

To link climate change to climate justice means understanding it as an issue of ethics, global equality, human rights, democratic accountability, participation and historical responsibility (Roosvall & Tegelberg, 2018). Fraser (2008) distinguishes between three matters of injustice pertinent for this purpose: economic injustice, concerning issues of (mal)distribution; cultural injustice, concerning issues of (mis)recognition (of identity/status); and political injustice, concerning political representation. Traditionally, justice theories have been limited to justice within states (Fraser, 2008) and climate justice has been discussed inter-nationally, adhering to nation-state logics. In a globalizing age, however, justice must be understood and applied beyond nation-state logics, encompassing intra-national and transnational geographical scales (Fraser, 2008; Roosvall & Tegelberg, 2018). Geographical scales can furthermore be isolated or combined. Regarding combination, we distinguish between *scalar transcendence* (Christensen, 2013) and *scalar integration* (Roosvall, 2019), where transcendence is here understood in a broad sense as going beyond one scale (to another). Transcendence, for example, shows how local problems can help solve global problems, while integration means more consistently entangling at least two scales, rather than using one to get to the other. When climate justice is related to migration, it is also notable that both being forced to migrate from your native lands and being hindered from migrating from a dangerous place—for example, due to lack of economic means—constitute forms of injustice.

Material and Method

This is an explorative pilot study of transnational and local reporting on climate migration, using the force of examples rather than claiming generalizability (Flyvbjerg, 2006). In each case, we chose three local news articles, and two articles from *The Guardian* that help us illustrate potential differences between local and transnational reporting.

The Guardian was selected because it is recognized for environmental and climate journalism, it has a transnational scope and is *not* invested in the national-geographical scale of the US. The news site was searched using terms for the three islands/communities, combined with the terms ‘climate’ and ‘migration/relocation’, and, in the case of Puerto Rico, the names of two major 2017 hurricanes. Long articles and articles focusing on the specific islands were prioritized over short articles and articles that were not specifically focused on the islands.

For the Alaska case, *Anchorage Daily News* (AD) and the *Nome Nugget* (NN) were identified as relevant local sources due to their proximity to Shishmaref. A Google search for the keywords ‘Shishmaref/Sarichef Island local news’ yielded one long article in AD and two long articles on the NN website.

For the Louisiana case, where recent news from the island has mainly focused on climate-induced relocation, a web search for “Isle de Jean Charles local news” identified relevant local news sources. Two articles from *Nola.com* and one from *The Advocate* were selected due to length and focus. *Nola.com* is an online site connected to *The Times Picayune*, a New Orleans-based print newspaper. *The Advocate* is the largest daily newspaper in Louisiana, based in Baton Rouge, serving the southern part of the state.

Relevant local sources were found in a similar way for Puerto Rico, with search terms added for 'Hurricane María/Hurricane Irma'. The longest articles available on relocation after Hurricane María were selected from *El Vocero*, *El Nuevo Día* and *Metro* respectively—three of the four daily newspapers with island-wide coverage currently published in Puerto Rico.

We explore whether people have 'a name, a face and a story' (Cavarero, 2002), drawing on the multimodal critical discourse analysis tools naming, visual representation and presupposition (Machin & Mayr, 2012). Regarding story, we thus consider not only how climate change, people, places and in/justice (economic/cultural/political) are explicitly addressed, as indicated in the RQs, but also what is presupposed—that is, implied as common sense.

Isle de Jean Charles, Louisiana

Isle de Jean Charles, Louisiana, is home to the Band of the Biloxi-Chitimacha-Choctaw Tribe, which ended up there after being displaced by 'Indian Removal Act-era policies' (isledejeancharles.com, n.d.). Since the 1950s, 98% of the island has vanished, with the remaining hundred or so residents planned to relocate in 2020. Coastal erosion was caused when oil and gas companies dredged canals and built pipelines that made saltwater destroy the former freshwater wetlands (Arizona State University, 2008). The ecosystem was destroyed further by attempts to protect the island with a levee. Moreover, sinking land, caused by lack of soil renewal, and increasing flood risk, combined with the effects of severe hurricanes, has already caused many to relocate, as has the need for access to reliable jobs and services (isledejeancharles.com, n.d.). The small schoolhouse was closed over 50 years ago; thus, many on the island lack formal education (Arizona State University, 2008). The tribe 'lacks federal recognition, making residents ineligible for federal assistance from the Federal Emergency Management Agency and the Bureau of Indian Affairs' (Arizona State University, 2008). According to tribal chief Albert Naquin, '[the Bureau of Indian Affairs] know we're Indians. We know we're Indians, but they just won't give us recognition because we don't have the proper historical records' (Ibid.). Yet, many inhabitants have been hesitant to move since the island holds the bones of their ancestors, and the fabric of their culture.

The Guardian's 'Louisiana's vanishing island: the climate "refugees" resettling for $52m'(15 March 2016) acknowledges climate change but uses quotation marks around 'refugees', indicating the term/status is not accepted. The people and the island are termed 'test case' and 'testing ground' for the resettlement of other communities. Hence, there is scalar transcendence, connecting local and global scales, but not scalar integration. That is, the community comes out as interesting mainly in how it can help others, not so much as *concurrently interesting for its own sake*. People on the island have names, faces and stories: there is a close-up of Wenseslaus Billiot 'an 88-year-old native', and a full body image of three named men. Additionally, there are four images of scenery and one of placards that tell a forceful story: 'Don't give up your rights.' This connects to justice more clearly than the article text. In the text, economic issues are evoked regarding resettlement costs; culture regarding views of the community as a 'culturally sensitive one'; political representation is not addressed, but exclusion from a 'master-plan' for the area comes up.

The second *Guardian* article (27 May 2017) also acknowledges climate change as a cause. The lead of this photojournalistic story says the 'Native American' community is among the nation's most vulnerable, citing 'a report by 13 US federal agencies'. It contains 12 images. People with names and faces appear in five photos/captions. Two photos include names, while the people are facing away from camera. Three include unnamed people. While this type of journalism invites readers to get visually close to named people, stories naturally get cut short. Thus, while we find out that Faye Danos is against the suggested relocation, we do not learn why. Issues of economic and cultural injustice are very loosely implicated, by speaking of 'traditional food and income'. Issues of political injustice are lacking.

In a local *Nola.com* article (20 December 2017) containing a photo gallery and a Google map of the area, sea-level rise is acknowledged as 'triggered by climate change'. People on the island are described as climate refugees, without quotation marks. Local climate change impacts on infrastructure are detailed, and a significant part of the text covers plans for relocation, concerning costs, geography, infrastructure, and so on. One aerial photo/map outlining the relocation area and two scenic images of the site constitute the entry point into the photo gallery, which furthermore contributes names and faces of residents. One caption says: 'Albert Naquin, chief of the Biloxi-Chitimacha-Choctaw Indian tribe, helped negotiate a $48 million grant through the U.S. Department of Housing and Urban Development to relocate the tribe to higher ground'.

Thereby redistribution and political participation are indicated, although there is no detailed information on the level or conditions of participation. Sandra 'Cookie' Naquin also appears in the text, hoping that her scattered family will be reunited in the settlement, thereby loosely evoking culture and identity.

Another *Nola.com* article (21 March 2018) also acknowledges climate change as a relocation cause and, furthermore, as the reason the community will receive federal assistance. The text details the geography, politics and economics of the relocation. It is illustrated with an aerial photo of the relocation area outlined in red. Sources are mostly agencies and officials, but also chief Albert Naquin, who is critical about the move occurring too late, as life on the island is dangerous. A *YouTube* video feature shows aerial footage of the relocation area. There are no images of people in the video or article, only the name and short critique offered by the chief. Mis/recognition is not addressed explicitly, but the purchase does indicate some form of recognition (the term "Native American" is also used). Redistribution is more clearly addressed and political *representation* is lacking. General political aspects are however foregrounded in the article, detailing political decisions and actions. There are links to previous articles, for example an extensive photo gallery from 2016, including many faces. The emphasis is on saving the community, rather than on how the community can save the rest of the world. Local and national scales are integrated, discussing the community as one of several vulnerable communities nationally.

The Advocate article (21 December 2018) also acknowledges climate change as the cause and mentions climate refugees without quotation marks. It includes three maps/map-like aerial photos, one scenery photo and one close-up of chief Naquin. Political aspects are included, detailing negotiations, deals and infrastructure. Cultural recognition of identity and status is alluded to by the chief, who is quoted saying he 'wants the community brought back together in one place like it used to be after being scattered by people moving away from the storm dangers'. More explicitly, Naquin 'worries about dilution of the tribe's culture'. In a statement by the head of the relocation project, cultural, political and economic aspects come together: 'We want to move the people on the island in such a way that the community can sustain itself', including attempts to attract businesses.

Shishmaref, Sarichef Island, Alaska

Located on Sarichef Island, a four-mile stretch of land on the west coast of Alaska in the Chukchi Sea, Shishmaref is home to a population of approximately 600 primarily Iñupiaq residents. The community has experienced six flooding disaster declarations since 1988 (Marino, 2015, p. 5). A deadly combination of flooding, coastal erosion and land loss, all exacerbated by a rapidly warming Arctic climate, will eventually force the community to relocate. Shishmaref is among a group of climate-threatened small island states and coastal communities that have become emblematic global news media symbols of the human impacts of climate change (Callison, 2017; Marino, 2015). International news media narratives have victimized Shishmaref's Iñupiaq residents, portraying them as a 'face of climate change' while ignoring the community's resilience in dealing with a deep historical legacy of racism, colonialism and structural violence (Marino, 2015, pp. 5, 11).

A 2016 article in *The Guardian* (18 August 2016) describes a special election held in Shishmaref to determine whether the community should relocate at an estimated cost of $180 million. The story focuses on the link between poverty and climate change emphasizing the 'steep price tag' residents will face (maldistribution) regardless of whether they decide 'to relocate or to stay and add environmental defences'. The stories of local mayor Harold Weyiouanna and Esau Sinnok describe damages that have already occurred, and the high economic costs associated with erecting shoreline defences.

A 2018 article features two Shishmaref residents, Dennis Davis and Percy Nayokpuk, explaining how hunters in coastal indigenous communities are adapting their practices in response to dramatic changes in hunting conditions and in the behaviour of prey caused by climate change (*The Guardian*, 2 March 2018). Davis describes how he uses a drone to 'analyse ice conditions' and help ensure safety as ice conditions become 'ever more variable'. He mentions the greater distances that hunters must now travel to find prey. The story displays an image of Davis standing adjacent to a snowmobile, watching a drone fly above a patch of sea ice. While both *Guardian* stories place emphasis on economic (maldistribution) and human costs associated with climate change and migration, limited emphasis is placed on political representation or recognition of rights and identity.

Local and regional news outlets have also dedicated coverage to Shishmaref, overlapping in terms of content but differing significantly in the emphasis they place on community-level responses. For example, an *Anchorage Daily News* article (19 August 2016) overlaps with *The Guardian's* emphasis on Shishmaref's relocation vote. However, the AD story focuses on community dimensions of the decision, such as potential sites for relocation, the ballot questions residents were asked and the importance of their role in the relocation process. Moreover, the fact that 'residents want to stay together in the area' and 'not scattered in towns and villages around Alaska' is highlighted.

Calls for a community-led relocation process are evident in two stories published in the *Nome Nugget* (NN) (22 July 2016). Esau Sinnok is the central name and face in an in-depth article about his prolific efforts to widen awareness of the connection between climate change and displacement, and the particular challenges Shishmaref faces (*Nome Nugget*, 22 July 2016). Sinnok touches upon the recognition of rights and identity, stating that he wants 'a secure future' not just for him but also for 'the next seven generations'. The author describes Sinnok's choice to wear 'a collared shirt and tie underneath his kuspuk' and 'introduce himself in Iñupiaq' on a visit to the White House. A photograph shows Sinnok standing in this clothing alongside other award recipients. Aspects of political representation are evoked in references to Sinnok's efforts at 'getting more youth involved' and engaging 'more people of colour in environmental conversations'.

Another NN article (26 May 2017) connects Shishmaref to other coastal and island communities facing climate-induced displacement, using it to exemplify challenges associated with climate-induced migration and to emphasize the importance of involving community members in relocation efforts. Robin Bronen is the main face and voice in this story, acknowledged for her continuing efforts to 'create and implement a community-led relocation policy'. A photograph captioned 'Erosion Effecting Alaska Villages' depicts Bronen giving a presentation on these efforts. The story evokes elements of maldistribution, noting that there is no 'federal framework' nor any 'federal funding allocated to this type of relocation'. Bronen's voice stresses the importance of ensuring that every decision concerning relocation is 'led by the community' (political representation), especially in light of the harmful legacy of 'government forced moves'. Scalar integration and not 'just' scalar transcendence is applied, since Shishmaref is understood as interesting *in itself* rather than mainly as an example. All of the local coverage addresses the need for more just recognition and political representation for Shishmaref in relocation processes.

Puerto Rico

Puerto Rico, an archipelago situated between the Caribbean Sea and the North Atlantic Ocean, is not a US state, but an unincorporated territory, controlled by the US Congress, where its citizens have no representative with full voting rights. Puerto Ricans cannot vote for the president of the US unless they are living in the mainland US. According to Philip Alston, UN Rapporteur on Extreme Poverty and Human Rights, the lack of political rights and poverty are inextricably linked: 'If it were a state, Puerto Rico would be the poorest state in the Union' (Alston, 2017). In September 2017, the archipelago was struck by two high-magnitude hurricanes, Irma and María. After visiting the island in December 2017, Alston stated: 'I listened to individuals in poverty and civil society organizations on how these natural disasters are just the latest in a series of bad news for Puerto Ricans/…/an economic crisis, a debt crisis, an austerity crisis and, arguably, a structural political crisis' (Ibid.). Ultimately, the hurricanes had a huge economic and human impact, leading to an increase in poverty and migration to the US (Rivera, 2018).

Two articles in *The Guardian* focus on Puerto Rican migration from a US perspective. The first understands Florida as the local area to which Puerto Ricans arrive, jobless and resourceless (*The Guardian*, 12 October 2017). Their migration, referred to as 'arrival', is presented as the product of a long-standing economic crisis combined with a one-time meteorological incident. The concern is how Florida could, and if it should, accommodate such migration. The problem is not how Puerto Rico will be rebuilt, but the impact of its displaced population on Florida's housing, education and employment systems, and on local politics. There is no discussion of climate change. Despite references to Puerto Rico's fragile situation prior to María, its effect on the archipelago's precarity is described as surprising. The article is illustrated with two stock photos showing Puerto Ricans arriving in Port Everglades, which show faces of non-named 'evacuees' who are not given a voice.

The other *Guardian* article (9 August 2018) focuses on Florida and Puerto Rico. Puerto Rico is portrayed as a disaster zone. A migrant young woman, Claudia Báez, recalls: 'It looked like a bomb had fallen/…/It was pure destruction.' Florida is portrayed as challenging for Puerto Rican migrants, partly due to inaction by the Trump administration, which, according to an interviewee, refused to provide adequate federal disaster assistance to "evacuees" (i.e. not refugees). The main problem is migrants' struggles to find adequate and affordable housing in Florida. Hurricane María is characterized as a destructive natural disaster, 'probably responsible for the death of 4600 people'. There is no discussion of climate change. While there are no specific references to class or ethnicity, the specific migrants portrayed in the article are characterized as members of an educated family with professionally employed parents, and a photo shows they are white. Oblique references indicate the crucial role of family networks and money in facilitating the transition from the archipelago to the mainland, but luck is ultimately highlighted as a main factor. Regarding political representation, Puerto Ricans migrating to Florida are described as mistreated by Trump's government and as 'a powerful force' capable of organizing and having an impact in midterm elections.

The three selected articles by Puerto Rican daily newspapers show some variation. The *El Vocero* article (10 November 2017) focuses on Puerto Rico and the US. Starting with a reference to periodical movements of Puerto Ricans between archipelago and mainland for varied reasons, the article focuses on how Hurricane María, as a particularly intense event, may change this periodicity. Puerto Rico's economic stagnation, and the lack of resources of (some of the) migrants, are discussed. There are no explicit references to ethnicity or class, but the migrants portrayed are said to have relatives (or friends) in the US who can host them. The article prioritizes the stories of elder women when quoting displaced Puerto Ricans. No references are made to the island's governance system. This absence, combined with the personalization of the hurricane's impact through individual stories, gives the impression that islanders live by themselves, with no political system in place administrating their everyday lives.

The *El Nuevo Día* article (29 September 2018) focuses on Puerto Rico's distress after the hurricane and on what is described as the ensuing rise in migration's volume and duration. There are no references to climate change. Migration is characterized as a challenge for Puerto Rico's future, rather than for Florida. The article reports on a census of migrants, presenting displaced Puerto Ricans not as people with names and faces but as figures, and stresses the importance of having rigorous and clear data about the extent of migration to the US after the hurricane, which is difficult to produce in the absence of well-established cross-mechanisms for tracking population movements. Although migration post-María is signalled as problematic, the article does not unpack or discuss its implications.

The *Metro* article (19 September 2018) 'María transformed central Florida' focuses on Puerto Rico and on Florida, defined as the state that received the largest share of migration after Hurricane María. There are no references to climate change. Puerto Rican migration is characterized as 'historical' and contextualized vis-à-vis migration from Latin American countries to the US, pointing to Puerto Rico's double status as a US territory and a Spanish-speaking country. There are no straightforward references to economy or class. Class, however, is presupposed where a couple of migrants with names and faces tell their stories about deciding to leave Puerto Rico when they could not reach the (private) paediatrician for their kids and had to take them to a (public) hospital's emergency ward instead. This middle-class couple seeks to avoid the conditions typical of a public hospital's emergency ward in the aftermath of a major crisis. Political representation is marked with explicit references to the importance of Puerto Rican migrants to Florida registering to vote and organizing to fight for their (minority) rights and those of their relatives left behind on the island.

Discussion and Conclusions

There are differences between the understandings of local areas in *The Guardian* and in local news media. In the Louisiana case, the local area is considered in much greater depth by local media, as evidenced in maps, photos and descriptions of the political process. Likewise, in the Alaskan case the connection between the local area and its residents is understood as more political compared to *The Guardian* material. In the Puerto Rico case, the same goes for one local article that emphasizes the importance of local political engagement to some degree. However, the *Guardian* mostly focuses on Florida, while there is a clearer concern with linking Puerto Rico and Florida in the local material.

Regarding connections between local and global aspects, *The Guardian* coverage shows geographical *scalar transcendence*, for instance, when the Louisiana case is described as a test case for the rest of the world, or when impacts on Florida are foregrounded in the case of Puerto Rico. This is also consistent with previous media portrayals of Shishmaref residents as the 'face of climate change' (Marino, 2015), and the overall portrayal of indigenous peoples as victim-heroes, hailed for helping save the rest of the world while their own lands vanish (Roosvall & Tegelberg, 2018). Local coverage of the Louisiana case shows instead more *scalar integration* where Isle de Jean Charles and the people who live there are portrayed as interesting in themselves rather than just as a test case. The global scale is however not always connected in this coverage.

A clear connection between climate change and migration is expressed in local and *Guardian* coverage of Louisiana and Alaska. The term "refugee" is used once in quotation marks in *The Guardian*, signalling that it is not a fully accepted term for the case in question. In the Puerto Rico case, neither *The Guardian* nor the local material mentions climate change. Terms like 'arrival' and 'evacuees' are used instead of 'migration' and 'refugee'. Missing in all the analysed material is the notion that the emissions affecting these residents have mainly been caused by people living elsewhere, who have benefitted from pollution while islanders have largely remained poor on their diminishing lands; in this sense climate injustice is not acknowledged explicitly (see Johnson, 2009) even though various aspects of injustice are implied.

The people migrating/about to migrate have names, faces and stories in most of the local and *Guardian* material, but the essence of the stories varies in terms of recognition, redistribution and political representation. Most of the coverage acknowledges economic aspects. These economic aspects are related to justice in local journalism in the Louisiana and Alaska cases when redistribution issues are connected to issues of political representation. Justice framing of one such aspect thus seems to facilitate justice framing of the other. While culture and ethnicity are evoked concerning indigenous peoples, this does not necessarily mean that misrecognition is addressed. For instance, the problems the Biloxi-Chitimacha-Choctaw people face in Louisiana because they are not properly recognized as indigenous are not brought up in the material. But recognition is implicated by calls for political action in the local material from Alaska, in one of the local Puerto Rico articles mentioning minority rights, and in acknowledgement of the cohesive importance of the connection to land and culture in some of the local Louisiana material. Yet again, justice aspects come out more strongly when related to other justice aspects. This integration of matters of justice (economic, cultural, political) overlaps notably to some degree with the aforementioned integration of geographical scales. Evoking one type of integration may thus facilitate evoking the other.

This exploration shows that local and transnational journalisms bring different issues to the table. Only focusing on one of them is not enough to get the full picture of climate migration, nor of the situations in disaster-affected localities. The transnational journalism of *The Guardian* tends to infuse global perspectives, which are needed in relation to climate change, but sometimes missing in local journalism. Local journalism seems however to attend to many other indispensable features: local geographic specificity, infrastructure and political processes; integration of matters of justice; and scalar integration. Thus, in local journalism a focus on the needs of people in disaster-affected areas is not motivated by what the rest of the world can learn. Rather, these are viewed as crucial factors in themselves. In this regard, local journalism treats local and regional communities more justly. This is most enhanced in the local cases rather than in the regional case.

The tendencies found here point to the need for further consideration of why climate change is accepted as a cause for migration in some areas and not in others, and of what this may have to do with class variations among local inhabitants, and with distribution, recognition and political representation more broadly, both in stories *and* in policies. The way Puerto Rico came out differently here compared to the Alaskan and Louisiana islands points in this direction.

News Articles Analysed

AP. (2017, November 10). Puertorriqueños llegan a EEUU tras huracán María' ('Puerto Ricans arrive to the US after Hurricane María'). *El Vocero*. Retrieved from https://www.elvocero.com/actualidad/puertorrique-os-llegan-a-eeuu-tras-hurac-n-mar-a/article_f82b6246-ae46-11e7-a6c2-4757981e2652.html.

Baurick, T. (2017, December 20). Here's where residents of sinking Isle de Jean Charles will relocate. *Nola.com*. Retrieved from https://www.nola.com/environment/index.ssf/2017/12/site_chosen_for_relocating_isl.html.

Delgado, J. A. (2018, September 29). El estimado de la emigración de boricuas después de María es de 160,000 (Puerto Rican migration after María estimated at 160,000). *El Nuevo Dia*. Retrieved from https://www.elnuevodia.com/noticias/locales/nota/elestimadodelaemigraciondeboricuasdespuesdemariaesde160000-2450019/.

Demer, L. (2016, August 19). Shishmaref votes to relocate from eroding barrier island to mainland. *Anchorage Daily News*. Retrieved from https://www.adn.com/alaska-news/2016/08/18/eroding-village-of-shishmaref-votes-in-favor-of-relocating-to-mainland-a-key-step/.

Holpuch, A. (2016, August 18). Alaskan village threatened by rising sea levels votes for costly relocation. *The Guardian*. Retrieved from https://www.theguardian.com/us-news/2016/aug/18/alaska-shishmaref-vote-move-coastal-erosion-rising-sea-levels.

Hughes, Z. (2018, March 2). In doomed Alaska town, hunters turn to drones and caribou as sea ice melts. *The Guardian*. Retrieved from https://www.theguardian.com/environment/2018/mar/02/alaska-climate-change-indigenous-hunting.

Levy, A./AP/Getty Images. (2017, May 27). Facing climate change on the Louisiana bayous—In pictures. *The Guardian*. Retrieved from https://www.theguardian.com/artanddesign/gallery/2017/may/27/climate-change-louisiana-bayous-isle-de-jean-charles.

Luscombe, R. (2017, October 12). Arrival of Puerto Ricans post-hurricane Maria could have big impact on Florida. Retrieved from https://www.theguardian.com/us-news/2017/oct/12/florida-puerto-rican-influx-hurricane-maria.

Luscombe, R. (2018, August 9). Exiled in Florida: The Puerto Ricans struggling to build a new life off island. Retrieved from https://www.theguardian.com/world/2018/aug/09/exiled-in-florida-puerto-ricans-struggling-to-build-a-new-life-off-island.

Mercado, D. C. (2018, September 19). María transformó la Florida central (María transformed central Florida). *Metro*. Retrieved from https://www.metro.pr/pr/noticias/2018/09/19/maria-transformo-la-florida-central.html.

Roberts, F. A. (2018, December 21). State chooses site near Thibodaux to relocate Isle de Jean Charles climate refugees. *The Advocate*. Retrieved from https://www.theadvocate.com/baton_rouge/news/environment/article_4c46b2ba-e685-11e7-9c86-0f1dd0c64526.html.

Sanolli, L. (2016, March 15). Louisiana's vanishing island: The climate "refugees" resettling for $52m. *The Guardian*. Retrieved from https://www.theguardian.com/environment/2016/mar/15/louisiana-isle-de-jean-charles-island-sea-level-resettlement.

Schleifstein, M. (2018, March 21). State is buying Isle de Jean Charles relocation site for $11.7 million. *Nola.com*. Retrieved from https://www.nola.com/environment/index.ssf/2018/03/state_is_buying_isle_de_jean_c.html.

Thomas, M. (2016, July 22). Shishmaref's Esau Sinnok recognized by White House as champion of change. *The Nome Nugget*. Retrieved from http://www.nomenugget.com/news/shishmarefs-esau-sinnok-recognized-white-house-champion-change.

References

Alston, P. (2017). Statement on visit to the USA. Retrieved January 9, 2019, from https://www.ohchr.org/EN/NewsEvents/Pages/DisplayNews.aspx?NewsID=22533&LangID=E.

Arizona State University. (2008). Tribes: Gulf coast. Biloxi-Chitimacha-Choctaw Indians: Rising tides. Retrieved from http://www7.nau.edu/itep/main/tcc/Tribes/gc_choctaw.

Baldwin, A. (2016). Resilience and race, or climate change and the uninsurable migrant: Towards an anthroporacial reading of race. *Resilience, 5*(2), 129–143.
Bardsley, D. K., & Hugo, G. J. (2010). Migration and climate change: Examining thresholds of change to guide effective adaptation. *Population and Environment, 32*(2–3), 238–262.
Betts, A. (2013). *Survival migration: Failed governance and the crisis of displacement*. Ithaca: Cornell University Press.
Callison, C. (2017). Climate change communication and indigenous publics. In H. von Storch (Ed.), *Oxford research encyclopedia of climate science*. Oxford: Oxford University Press.
Cavarero, A. (2002). *Stately bodies: Literature, philosophy, and the question of gender*. Ann Arbor, MI: University of Michigan Press.
Christensen, M. (2013). Arctic climate change and the media. In M. Christensen, A. E. Nilsson, & N. Wormbs (Eds.), *Media and the politics of Arctic climate change* (pp. 26–51). Basingstoke: Palgrave Macmillan.
Dreher, T., & Voyer, M. (2015). Climate refugees or migrants? Contesting media frames on climate justice in the Pacific. *Environmental Communication, 9*(1), 58–76.
Farbotko, C., & Lazrus, H. (2012). The first climate refugees? Contesting global narratives of climate change in Tuvalu. *Global Environmental Change, 22*(2), 382–390.
Flyvbjerg, B. (2006). Five misunderstandings about case-study research. *Qualitative Inquiry, 12*(2), 219–245.
Fraser, N. (2008). *Scales of justice: Reimagining political space in a globalizing world*. Columbia University Press.
isledejeancharles.com. (n.d.). Bienvenue, Aiokpanchi, welcome to Isle de Jean Charles. Retrieved from http://www.isledejeancharles.com/.
Johnson, S. (2009). Climate change and global justice: Crafting fair solutions for nations and peoples. *Harvard Environmental Law Review, 33*(2), 297–301.
Machin, D., & Mayr, A. (2012). *How to do critical discourse analysis*. London: Sage.
Marino, E. (2015). *Fierce climate sacred ground: An ethnography of climate change in Shishmaref, Alaska*. Alaska: Alaska University Press.
Piguet, E., Pécoud, A., & de Guchteneire, P. (2011). Migration and climate change: An overview. *Refugee Survey Quarterly, 30*(3), 1–23.
Rivera, M. (2018). The crisis before the crisis in Puerto Rico. Retrieved January 9, 2019, from https://socialistworker.org/2018/01/16/the-crisis-before-the-crisis-in-puerto-rico.
Roosvall, A. (2019). Här och där: För en humanistisk journalistisk geografi i globaliseringens och klimatförändringarnas tid [Here and there: For a humanistic, journalistic geography in times of globalisation and climate change]. In *Humanistiska fakulteten 1919–2019: Nedslag i humaniora*. Stockholm: Stockholm University.
Roosvall, A., & Tegelberg, M. (2018). *Media and transnational climate justice: Indigenous activism and climate politics*. New York: Peter Lang.

PART II

Armed Conflict and Journalistic Freedoms

CHAPTER 7

Changing the Story of Urban Violence in El Salvador: The *Crónica*, the Community, and Voices from the Ganglands

Mathew Charles

Violence or the fear of violence is a daily reality for millions of people around the world. Some of the highest homicide rates occur in cities that are not actually at war. For example, between 1978 and 2000, more people (49,913) were killed in the slums of Rio de Janeiro than in all of Colombia, a country experiencing civil conflict (Dowdney, 2004). Indeed, Latin America and the Caribbean (LAC) remains the world's most violent region, according to the Mexican NGO the Citizen Council for Public Security and Criminal Justice (CCSPJP). The annual report from the CCSPJP concluded that 43 of the world's most dangerous cities in 2018 were in the LAC region, with the remainder in the USA and South Africa.

In El Salvador, there were 3340 homicides in 2018, making it one of the most murderous countries in the world. The official murder rate of 51 per 100,000 inhabitants is second only to Venezuela in Latin America and the Caribbean (Insight Crime, 2018). The United Nations Office on Drugs and Crime (UNODC) says that 30% of the homicides in the

M. Charles (✉)
Bogota, Colombia

J. Matthews, E. Thorsen (eds.), *Media, Journalism and Disaster Communities*, https://doi.org/10.1007/978-3-030-33712-4_7

Americas are attributed to organised crime and gang-related violence. This is compared to less than 1% in Asia, Europe, and Oceania (UNODC, 2013).

El Salvador's violence has long been linked to the bitter rivalry between the country's biggest gangs: the *Mara Salvatrucha* (MS-13) and the two factions of *Barrio 18*. They control marginalised communities across the country, where residents are held hostage by "invisible borders", which define gang turf. Trapped in their neighbourhoods, citizens are unable to move freely as crossing into rival territory nearly always carries a death sentence. Small and large business owners become victims of elaborate extortion rackets, forced to pay monthly fees or face execution.

Since its inception, the online news portal, *El Faro* or The Lighthouse, has set out to change the story on El Salvador's gangs, predominantly by incorporating the voices of gangsters into their coverage. Urban violence is all too often covered from the "official" perspective. That is to say through government statistics and interviews with the authorities, including the police and military, which does not necessarily produce an accurate or balanced picture of how communities living with violence on the ground may experience gang conflict. By talking to those who perpetrate violence and those who must confront it on a daily basis within their neighbourhoods, journalists may discover truths, which question and counter dominant arguments, perhaps even exposing state and institutional failures.

This chapter therefore argues that a more inclusive journalism reconstitutes the notion of community to emphasise the importance of human connection. First, the chapter explores how urban violence perpetrated by gangs damages the social fabric of community and charts how the dominant media coverage of gangs only serves to perpetuate fear, failing to acknowledge the complexities of what are now largely considered to be a sophisticated social phenomenon.

Urban Violence: Breaking the Community

The United Nations Office on Drugs and Crime (2013) says it is the poorest communities who suffer most from gang violence. In areas affected by high levels of violence, life becomes chaotic, uncertain and citizens become powerless. The violence pushes people to the edge of their emotional and physical strength (Nordstrom & Robben, 1995) and potentially even towards violence themselves (Romero, 2003, p. 17; Vásquez, 2006, p. 341). Since violence can come from anywhere for whatever reason, all citizens become "potential victims" (Rotker, 2002, p. 15). As a result, nobody from beyond the intimate circle of family and friends can be trusted. Public institutions meant to provide security and protect citizens' rights are non-performing, absent, or have become part of the threat. Susan Rotker (2002) and Jésus Martín-Barbero (2002) therefore portray fear as a permanent attribute of the Latin American citizen.

This devastation of the social fabric creates mistrust and halts social interactions (Clancy & Hamber, 2008; Herman, 2015; Lykes & Mersky, 2006). It instigates "collective trauma" (Abramowitz, 2005) and "social suffering" (Kleinman, Das, & Lock, 1997).

In communities with a strong presence of gangs, the state often increases its mechanisms of control in an attempt to counteract their growing power and influence. Therefore, as Winton (2004) argues, these communities are subject to manipulation by the state, the gangs, and the elite political sector, and are permanently caught between multiple power systems.

The highly visible nature of gangs and urban violence can result in a phenomenal amount of media attention. Yet media reports can exaggerate the actual occurrence of gang violence and often serve "to obscure far more insidious violence" (Moser & Winton, 2002, p. 25).

Caldeira and Holston (1999) have argued that urban violence in Brazil generates a "talk of crime" in which the repetition of stories about crime produces feelings of fear. This spiral of fear and violence, Caldeira and Holston argue, has a profoundly negative impact, producing a "disjunctive democracy", where rights such as freedom of expression and assembly are largely honoured, yet rights such as freedom from torture, summary execution or arbitrary arrest are routinely violated. Fear, in turn, leads people to restrict their movements and exacerbates violence by encouraging illegal responses to perceived criminality, such as supporting death squads and violent policing (Penglase, 2007).

Penglase (2007) concludes that gangs and violence are represented by images and metaphors that focus on the danger of "infection" and which advocate the need for "prophylaxis". Macek (2006) has also identified a "discourse of savagery" by which the media vilify urban youth and deploy metaphors of "contagion" and "penetration", often delivering grisly details in a sensationalistic tone. As Reisman (2006) has pointed out, this kind of sensationalistic media coverage, "often displaying images of tattooed young men being arrested or bloody shots of injured victims and corpses", has "contributed to a culture of fear that encourages government suppression, with little public support for a more balanced approach" (2006, p. 150). Esbensen and Tusinski (2007) assert that this sensationalism is based on a stereotypical representation of gangs. They argue that youth gangs are problematic enough in reality without the media contributing to exaggerations of their attributes.

Similarly, Correa Cabrera (2012) explains how the mass media have created what he calls a "spectacle" of border violence in the U.S. as an offshoot of Mexico's drugs war. This perpetuates an alarmist view, which projects an inaccurate "spillover" of violence into the United States.

Gunckel (2007) describes how what he calls "gangsploitation" documentaries foster "an exploitative gaze", which depicts gang life as entertainment. This glorification of gang membership and gang culture is sometimes apparent even through the "sobriety" of print and television journalism, he argues (2007, p. 38). For Gunckel, "gangsploitation" includes the perpetual use of "sensationalistic imagery" and emphasises "intensified enforcement", rather than examining structural causes or proposing a broader range of solutions to the violence (2007, p. 38). This argument is supported by Poynting, Noble, and Tabar (2001), whose study of Middle Eastern gangs in Sydney, Australia, concludes that complex class related social realities are over-simplified, misread, and misrepresented.

Huhn, Oettler, and Peetz (2006) argue that it has become common to state that criminal violence has superseded political violence in Central America to the extent that it has become normalised. There are signs that the problem of juvenile delinquency is emerging as "ordinary violence" in the region. They argue that the leading newspapers in Costa Rica, El Salvador, and Nicaragua tend to neglect critical news related to the deeper roots of criminal violence. That is to say that gangs and organised crime simply become part of the discourse to such an extent that it becomes an accepted characteristic of social life. What we might therefore call irresponsible coverage of gangs serves to fuel gang violence, rather than question its causes.

In contrast, a more responsible journalism about urban violence should acknowledge that the phenomenon is rooted in a range of exclusions, which range from the macro to the micro (Winton, 2004). At the macro level, the state fails to make citizenship meaningful. There is perhaps a lack of education, for example. As a result, there is increasing social fragmentation and polarisation. Furthermore, youth exclusion has also been related to globalisation. That is to say that the gangs, through their engagement in crime, offer the prospect of economic sufficiency in the face of a lack of alternative opportunities (Cerbino, 2012; Cruz, 2011). Similarly, at the micro level, research points to the absence of family cohesion and support. Gang membership can therefore provide a sense of belonging and community (Cerbino, 2012; Rodgers & Muggah, 2009).

The limited analysis of gangs is not just to be found in journalism. As Cruz (2011) has noted, most studies on urban violence tend to focus on organised crime and the "criminalisation of youth" (2011, p. 21), without considering that the states themselves "foster the conditions for the escalation of violence" (2011, p. 8). The blatant and instrumental "criminalisation" of gangs by the state has become a means of concealing more fundamental social and economic injustices. Rodgers and Muggah (2009, p. 12) have suggested that gangs have become a "scapegoat" onto which the state pins blame for its own failings, an idea supported by the scholar Mauricio Cerbino (2012). Cruz (2011) also suggests that violent crackdowns on crime have deflected people's attention from the corruption of officials and the state's inadequacies.

El Faro has built up its reputation with "stirring stories of gangland, sparking threats at home and racking up prizes abroad" since 1998 (Maslin, 2016). With more than a million unique visitors per month, it has become one of Latin America's most trusted news sites. In 2011, El Faro launched an investigative section of its website with five reporters, two photojournalists and a documentary filmmaker. The *Sala Negra* or the Black Room was created to specifically investigate the country's gang phenomenon. Its reporters ventured deep into the lion's den of gang turf and into what had previously been uncharted territory for El Salvadoran journalists. Their journeys into the criminal underworld shocked and impressed. Their access to gangs and their members was unprecedented. As Maslin (2016) has pointed out, the *Sala Negra* team began to ask hundreds of simple questions: Why did gangsters tattoo their faces? Why did teenage recruits accept being brutally beaten in order to become official members? Why did gangs extort poor businesses? But the more time they spent in El Salvador's conflict affected communities, the more they began to understand and the more they began to challenge the dominant narrative on gangs. In an interview with the author, El Faro's editor, José Luis Sanz, said that their goal was simple:

> We set out to understand gang violence from those who perpetrate it and those that live with it. We suspected that the root cause of the gangs were

> founded in social issues and that's what we found. This mattered because it highlights the failures of the state and contradicts the government's narrative that these guys are just criminals who deserve punishment. We wanted to go beyond coverage of body counts. Our aim was to break away from the over-simplified narrative of good versus evil.

Changing the Story: Voices from the Ganglands

It is easy to demonise those who perpetrate crime, but not as easy to question their motives or identify the structures, which may be underpinning the wider issue of urban violence. By deconstructing the victim/perpetrator dichotomy through the inclusion of a wider range of voices, reporters can explore and point to long-term structural issues, which may in turn promote social cohesion and provide a deeper contribution to our understanding of urban violence and gangs.

When they launched, the *Sala Negra* team was determined that their journalism would be different to the "mainstream", not just in terms of who they spoke to, but how they presented their stories. They settled on a long-form, narrative journalism, which would "provide the space needed to explore a complex issue in detail" (José Luis Sanz in an interview with the author in 2016).

Known as the *crónica*, this literary style of reporting comes from a great tradition in Latin America, where it is often associated with writers such as the Colombian Gabriel Garcia Marquez or the Brazilian Carlos Drummond de Andrade. It is a popular genre across the continent and a regular feature of newspapers and the journalism school curriculum. As journalist Gerardo Reyes told Silvio Waisbord (2000) in 1995:

> Latin American journalism accepts literary license and sacrifices facticity. It paraphrases rather than cites sources. It doesn't believe in quotes. It interprets without facts. (Reyes, 1995 in Waisbord, 2000, p. 140)

In Colombia, this style has been able to more easily influence public opinion, because it is "more carefully read" (Arroyave & Barrios, 2012, p. 400), but this narrative genre requires a distinct and immersive form of newsgathering or data collection.

Immersion in the contemporary era of journalism studies tends to refer to a new journalism production such as virtual reality, which creates a first person experience of a particular situation. Immersive technologies have been used to conceive a sense of "being there" (Jones, 2017), but technological developments aside, this concept of "being there" is not as new as it is sometimes presented. Indeed, this chapter argues that the literary journalism of the crónica is also immersive because it requires the reporter to spend long periods of time in the field getting to know his or her contributors through close relationships, but additionally because it fosters a more intimate bond with the reader, who is pulled into the world of the story through narrative construction. The crónica blends the stylistic art of storytelling and its conventions of fiction with the real world of gathering, interpreting, and researching information that is associated with journalism:

> This gives us the time and space we need to explore and explain a complex issue in detail. By presenting gangsters as real people in real situations, the aim becomes to help other people understand their lives. This form of reporting allows us to humanise the issue. (José Luis Sanz in an interview with the author in 2016)

The journalists at El Faro say their initial access to the normally well-hidden gang population was largely due to the time they spent building trust and to the particular strategy they devised for starting with the imprisoned gang leadership. In an attempt to win over journalists during the truce period of 2012 when MS-13 and Barrio 18 decided to stop killing each other in return for better conditions behind bars, the government had given El Faro unprecedented access to the country′s prisons. The gangs had then followed suit in the communities they controlled. Sala Negra emerged with a deeper relationship to the gangs:

> We started to work with leaders in the prisons. This led to them putting us in touch with other gangsters on the streets. There is no magic formula but together we decided that before we spoke to anyone in the barrios, we would do our research first. So we looked into their structures and their customs. We didn′t want to come across as naïve. (José Luis Sanz in an interview with the author in 2016)

After initial contact, it is often simply a case of the "snowball" effect, whereby one gang member introduces the next, and the next, and so on, but "it is not easy, nor is it guaranteed" (José Luis Sanz in an interview with the author in 2016). The El Faro team had shown that they were serious by carrying out research, which demonstrated to potential contributors that they were well-informed and unlikely to exploit them or hand them over to the security forces.

The Sala Negra team discuss ethical issues generated by their interviews as part of their daily routine. For example, one of their first dilemmas was how to handle the confession of criminal activity, including murder, but the team is adamant that they "are not there to do the work of the police or a judge" (José Luis Sanz in an interview with the author in 2016). The El Faro reporters have gained more trust overtime because gangsters have learned that contributions to the Sala Negra's reporting have been confidential.

Many of El Faro's sources have emerged from the large groups of *criteriados* or protected witnesses within El Salvador's justice system. These are usually gangsters who have been involved in committing a crime, but in exchange for their testimony they are not given a custodial sentence and are given protective measures until the case or cases in which they are involved are complete. Working with vulnerable sources is of course challenging on many fronts, but being interviewed also seemed to generate a sense of empowerment among many gangsters:

> The first rule of interviewing is to listen. And just by listening you can earn trust. Many of these gangsters had never been listened to before, not like this. And they felt good about it. (José Luis Sanz in an interview with the author in 2016)

The "politics of listening" is of course a key aspect of inclusion (Dreher, 2010; Threadgold, 2006). In her analysis of asylum issues, Terry Threadgold asserts that the absence of a "minority discourse" or counter-narrative is a result of "a lack of dialogue which might recognise the story of the 'other' as worth telling or hearing/seeing" (2006, p. 233). For Dreher, entrenched news values and existing story agendas focus on "the stereotypes and concerns of a mainstream audience rather than providing an open forum for speaking up" (2010, p. 99). By including the voices of gangsters and perpetrators of crime, the aim of El Faro has been to provide a more complex narrative of the urban violence and gang conflict in El Salvador. It has also been to generate a more informed debate about dominant security policy. Across Central America, where about 95% of crimes go unpunished in some areas, the so-called mano dura or iron fist policies have been pursued since the early 2000s in efforts to curb gang-fuelled violence (Hume, 2007). In El Salvador, for example, anyone can be arrested and charged with belonging to a gang just for having a tattoo. The iron fist approach has been widely criticised for human rights abuses and for failing to achieve its goals of a reduction in homicide and other violent crime:

> With mano dura police officers became judges in situ. Nobody was talking about the impact these hardline policies were having on local communities, which effectively became police states. (José Luis Sanz in an interview with the author in 2016)

The journalism of El Faro and its Sala Negra team is therefore attempting to produce a journalism that breaks away from the dominant corporate structures of so-called mainstream news and construct new and more inclusive narratives of gang violence. This would perhaps be characterised by Dreher (2010, p. 88) as a "community media intervention", where the aim is to deliberately challenge the mainstream and "expand, diversify or contest the range and types of representation" in order "to produce and mobilise counter narratives".

The work of El Faro has of course not been without risk, but, interestingly, most threats to El Faro reporters emerge from state actors, rather than the gangs. The founder of El Faro, Carlos Dada, says he and his team are regularly followed and photographed by intelligence agents. Indeed anti-press violence is not rare in El Salvador. The whole of the editorial team received death threats in 2012 after revealing that the government had held secret talks with gangs in the country. In an interview with the author, Dada explained how despite acknowledging that the El Faro journalists were in danger, the country's security minister at the time, David Munguía Payés, refused to offer them protection (Dada, 2016). In 2017, further threats were received in response to an article about a specialist anti-crime unit and their alleged involvement in criminal activity (CPJ, 2017).

According to Silvio Waisbord (2002, p. 92), anti-press violence in Latin America is a definitive outcome of state failure because, in Weberian terms, it implies the loss of its monopoly over violence. For Rockwell and Janus (2003), a weak media pluralism, where "the mainstream press and the state have consistently been close" (Waisbord, 2007, p. 309) is a defining structural characteristic of what they call "new democracies" or post-authoritarian states (see also Voltmer, 2013). In El Salvador, a former military dictatorship, which emerged from 12 years of brutal civil war in 1992, journalistic autonomy from the political and economic sectors is "inconceivable when newsrooms have a limited degree of independence vis-a-vis owners' economic and political interests" (Waisbord, 1996, p. 356). It was against this backdrop that the independent El Faro emerged:

> We wanted to produce a journalism that was more useful to society and not just based on news, a journalism that was not tied to the elite but one that could hold them to account. Journalism is not supposed to support the government agenda, but to challenge it. Our political system has been hijacked by the two extremes of left and right and we need to show alternatives to this polarisation. (Carlos Dada in an interview with the author in 2016)

El Faro therefore rejects a system of news, which they believe serves the interests of the political and economic elites at the expense of the so-called voiceless. If mainstream reporters are the "gatekeepers", who craft and control what is being published to the masses, then the Sala Negra reporters are the "gatebreakers", fighting to include a wider range of voices and interests in their coverage. By including voices of the marginalised, the journalists provide a space in which those who live on the edges of society can participate more effectively in social life and voice a perspective, which might otherwise be ignored. By asking their readers to listen, such inclusion encourages a greater social cohesion and ultimately attempts to foster a deeper sense of understanding and community or connectedness.

Conclusion: Re-connecting the Community

The Sala Negra's knowledge of the gangs is unrivalled in the journalism world and often surpasses that of academics and law enforcement. For Carlos Dada, this is why El Faro has been able to build the loyal readership he describes as a "community". Community journalism is usually conceptualised by a "nearness to people" (Byerly, 1961 cited in Robinson, 2013), but changes in media technologies have led to a reiteration of community journalism as a geographic-based phenomenon (Reader & Hatcher, 2012). As Robinson (2013) notes, at the contemporary "intersection of community, news, and digital technologies", we are questioning not only our news modalities, but also "our very understandings of what is community" (2013, p. 114). Friedland (2012) makes the case for a more virtual consideration of community journalism, which connects citizens not through geography, but through particular or special interests. The danger of such "alternative journalism" (Atton & Hamilton, 2008), often created to counter monolithic and generic forms of journalism, is their potential to engender "filter bubbles", which can prohibit a diverse exchange of views and promote or exacerbate polarisation (Flaxman, Goel, & Rao, 2016). That they are preaching only to the converted is an accusation often thrown at El Faro. Sanz (2016) acknowledges that the response to their work has not always been positive, but he believes this is a direct result of the country's extreme social and political divisions and is evidence of how mainstream corporate media have been able to successfully project a "false narrative" of the gang conflict:

> Sometimes the response to our work has been aggressive but that's because the dominant narrative remains strong. If we don't present stories from the perspective of the gangsters then we will never understand the true nature of this violence and we will never break the polarisation that exists. (José Luis Sanz in an interview with the author in 2016)

In post-conflict societies like that of El Salvador, conflict and human rights violations have "an enduring impact on individuals and communities" (Nickerson, Bryant, & Litz, 2012, p. 9). As Newman and Nelson (2012) have noted in their work on trauma in contexts of post-conflict, journalists who produce sensationalist and over-simplified stories risk increasing social "fragmentation". Conversely, a journalism that covers the causes and solutions of conflict by including all parties involved can lead to social "integration":

> At the end of the day our aim was to be more inclusive, but also more intelligent. We wanted to treat our readers with more intelligence and present them with more complex narratives away from the consequences of crime to the causes of it in the first place—to think of the gang phenomenon as a social issue as opposed to a criminal one. (Carlos Dada in an interview with the author in 2016)

The journalism of El Faro therefore constitutes the reconnection of community through its inclusive storytelling and its challenge to the dominant representation (and perhaps even exclusion) of the country's marginalised populations. Through their inclusion of diverse voices and specifically those of gangsters, El Faro journalists become the facilitators of an evanescent "public sphere" (Habermas, 1989), or perhaps more appropriately, a "civil sphere", which "stresses the critical role of social solidarity" (Alexander, 2006, p. 43). By facilitating a civil sphere, these journalists are attempting to repair the social bonds broken by gang violence, while fostering a deeper understanding of that violence and its underlying causes. Reader (2012) has emphasised that for community journalists to be successful today, they would nurture a particular connection to their audiences. In other words, journalists should report and write thinking about the citizens, as opposed to the place. Community in this sense relies not on geography, but instead on human connectedness. It is within and through these connections that people can begin to make sense of their lives and journalists can support and sustain the community and what it means to belong.

References

Abramowitz, S. A. (2005). The poor have become rich, and the rich have become poor: Collective trauma in the Guinean Languette. *Social Science and Medicine, 61*(10), 2106–2118.

Alexander, J. (2006). *The civil sphere*. Oxford: Oxford University Press.

Arroyave, J., & Barrios, M. (2012). Journalists in Colombia. In D. Weaver & L. Willnat (Eds.), *Journalists in the 21st century* (pp. 400–412). New York: Routledge.

Atton, C., & Hamilton, J. F. (2008). *Alternative journalism*. London: Sage.

Caldeira, T., & Holston, J. (1999). Democracy and violence in Brazil. *Comparative Studies in Society and History, 41*(4), 691–729.

Cerbino, M. (2012). *El Lugar de la Violencia. Perspectivas Criticas Sobre Pandillerismo Juvenil*. Quito: FLACSO.

Citizen Council for Public Security and Criminal Justice (CCSPJP). (2018). *Seguridad, Justicia y Paz*. Mexico City: CCSPJP.

Clancy, M. A. C., & Hamber, B. (2008). *Trauma, peacebuilding and development: An overview of key positions and critical questions*. Paper presented at the Trauma, Development and Peacebuilding Conference, Delhi, September 2009. Retrieved December 1, 2015, from http://www.incore.ulst.ac.uk/cgi-bin/projects.pl?task=fullout&id=43.

Correa-Cabrera, G. (2012). The spectacle of drug violence: American public discourse, media and border enforcement in the Texas-Tamaulipas border region during drug war times. *Norteamérica, 7*(2), 199–220.

CPJ. (2017). Two Salvadoran news websites threatened for their reporting. New York: Committee to Protect Journalists

Cruz, J. M. (2011). Criminal violence and democratisation in Central America: The survival of the violent state. *Latin American Politics and Society, 53*(4), 1–33.

Dowdney, L. (2004). *Neither war nor peace: International comparisons of children and youth in organized armed violence*. Rio de Janeiro: Viva Rio/Instituto de Estudos da Religiao.

Dreher, T. (2010). Speaking up or being heard? Community media interventions and the politics of listening. *Media, Culture and Society, 32*(2), 85–103.

Esbensen, F. A., & Tusinski, K. E. (2007). Youth gangs in the print media. *Journal of Criminal Justice and Popular Culture, 14*(1), 21–38.

Flaxman, S., Goel, S., & Rao, J. M. (2016). Filter bubbles, echo chambers and online news consumption. *Public Opinion Quarterly, 80*(1), 298–320.

Friedland, L. (2012). Community journalism at metropolitan ecology. In B. Reader & J. Hatcher (Eds.), *Foundations of community journalism* (pp. 151–153). London: Sage.

Gunckel, C. (2007). Gangs gone wild. Low budget gang documentaries and the aesthetics of exploitation. *The Velvet Light Trap, 60*, 37–46.

Habermas, J. (1989). *The structural transformation of the public sphere: An inquiry into a category of Bourgeois society*. Cambridge: Polity.

Herman, J. (2015). *Trauma and recovery: From domestic abuse to political terror* (3rd ed.). New York: Basic Books.

Huhn, S., Oettler, A. & Peetz, P. (2006). *Exploding crime? Topic management in Central American newspapers*. Working Paper No. 33. The German Institute of Global and Area Studies.

Insight Crime. (2018). 2018 homicide round-up. Retrieved from https://www.insightcrime.org/news/analysis/insight-crime-2018-homicide-roundup/.

Jones, S. (2017). Disrupting the narrative: Immersive journalism in virtual reality. *Journal of Media Practice, 18*(2–3), 171–185.

Kleinman, A., Das, V., & Lock, M. M. (Eds.). (1997). *Social suffering*. Berkeley: University of California Press.

Lykes, M. B., & Mersky, M. (2006). Reparations and mental health: Psychosocial interventions towards healing, human agency, and rethreading social realities. In P. De Greiff (Ed.), *The handbook of reparations*. Oxford: Oxford University Press.

Macek, S. (2006). *Urban nightmares: The media, the right, and the moral panic over the city*. Minneapolis: University of Minnesota Press.

Martín-Barbero, J. (2002). The city: Between fear and the media. In S. Rotker (Ed.), *Citizens of fear. Urban violence in Latin America* (pp. 25–36). New Brunswick, NJ: Rutgers University Press.

Maslin, S. E. (2016). A light in the underworld: Salvadoran website El Faro has built its reputation with stirring stories of gangland, sparking threats at home and racking up prizes abroad. *New York: Columbia Journalism Review*. Retrieved from https://www.cjr.org/special_report/el_faro_el_salvador.php.

Moser, C. & Winton, A. (2002). *Violence in the Central American region: Towards an integrated framework for violence reduction*. ODI Working Paper No. 171. Overseas Development Institute, London.

Newman, E., & Nelson, S. (2012). Reporting on resilience and recovery in the face of disaster and crime: Research and training implications. *Australian Journalism Review, 34*(1), 17–32.

Nickerson, A., Bryant, R. A., & Litz, B. T. (2012). Mechanisms of psychological harm following human rights violations and torture. *Stress Points, 26*(5), 9–11. International Society for Traumatic Stress Studies.

Nordstrom, C., & Robben, A. C. G. M. (1995). *Fieldwork under fire*. London: University of California Press.

Penglase, B. (2007). Barbarians on the beach: Media narratives of violence in Rio de Janeiro, Brazil. *Media, Crime and Society, 3*, 305–324.

Poynting, S., Noble, G., & Tabar, P. (2001). Middle Eastern appearances: "Ethnic gangs", moral panic and media framing. *Australian and New Zealand Journal of Criminology, 34*, 66–90.

Reader, B. (2012). Community journalism: A concept of connectedness. In B. Reader & J. Hatcher (Eds.), *Foundations of community journalism* (pp. 3–20). London: Sage.

Reader, B., & Hatcher, J. A. (2012). *Foundations of community journalism*. London: Sage.

Reisman, L. (2006). Breaking the vicious cycle: Responding to Central American youth gang violence. *SAIS Review, 26*(2), 147–152.

Robinson, S. (2013). Community journalism amidst media revolution. *Journalism Practice, 8*(2), 113–120.

Rockwell, R., & Janus, N. (2003). *Media power in Central America*. Chicago: University of Illinois Press.

Rodgers, D., & Muggah, R. (2009). Gangs as non-state armed groups: The Central American case. *Contemporary Security Policy, 30*(2), 301–317.

Romero, M. (2003). Paramilitary groups in contemporary Colombia. In D. Davis & A. W. Pereira (Eds.), *Irregular armies and their role in politics and state formation*. Cambridge: Cambridge University Press.

Rotker, S. (2002). *Citizens of fear. Urban violence in Latin America*. New Brunswick, NJ: Rutgers University Press.

Threadgold, T. (2006). Dialogism, voice and global contexts. *Australian Feminist Studies, 21*(50), 223–244.

United Nations Office on Drugs and Crime (UNODC). (2013). Global study on homicide. Vienna: UNODC. Retrieved from https://www.unodc.org/documents/gsh/pdfs/2014_GLOBAL_HOMICIDE_BOOK_web.pdf.

Vásquez, T. (2006). Dinámicas, tendencias e interacciones de los actores armados en el Magdalena Medio 1990–2001. In M. Archila & I. Bolívar (Eds.), *Conflictos, Poderes e Identidades en el Magdalena Medio 1990–2001* (pp. 313–372). Bogotá: CINEP.

Voltmer, K. (2013). *The media in transitional democracies*. Cambridge: Polity.

Waisbord, S. (1996). Investigative journalism and political accountability in South American democracies. *Critical Studies in Media Communication, 13*(4), 343–363.

Waisbord, S. (2000). *Watchdog journalism in South America: News, accountability and democracy*. New York: Columbia University Press.

Waisbord, S. (2002). Anti-press violence and a crisis of the state. *The International Journal of Press/Politics, 7*(3), 90–109.

Waisbord, S. (2007). Democratic journalism and statelessness. *Political Communication, 24*(2), 115–129.

Winton, W. (2004). Urban violence: A guide to the literature. *Environment and Urbanization, 16*(2), 165–184.

CHAPTER 8

Oscillating Between Alienation and Frustrated Engagement: The Study of Donbas Residents' Response to Conflicting Narratives in the Media

Dariya Orlova

2014 brought crucial changes and challenges to Ukraine. Mass protests, which became known as Euromaidan Revolution, regime change, the annexation of Crimea by Russia and the outbreak of armed conflict in the eastern region of Donbas were major disruptions that critically reshaped Ukrainian political life and social reality. The escalation of the conflict in Donbas caused significant loss of life and triggered a humanitarian crisis, which has remained acute despite a series of ceasefires. According to UN estimates, as of early 2019, 3.5 million people are described as being in need of "urgent humanitarian aid and protection" (UN OCHA, 2019a). Violent clashes, shelling and landmines, damaged infrastructure and disruption to essential services such as electricity and water are everyday realities for people living in Donbas.

D. Orlova (✉)
Mohyla School of Journalism, National University
of Kyiv-Mohyla Academy, Kyiv, Ukraine

J. Matthews, E. Thorsen (eds.), *Media, Journalism and Disaster Communities*, https://doi.org/10.1007/978-3-030-33712-4_8

The conflict in east Ukraine has also been characterized by Russia's efforts to spread propaganda. The communities of conflict-torn Donbas have thus suffered from the direct impacts of armed conflict and the vulnerabilities it creates or exacerbates, such as economic insecurity, environmental degradation and marginalization, but also faced a concerted campaign of disinformation. Contradictory news, emotionally charged claims and conflicting interpretations of developments in the conflict have been a feature of media discourse, one that audiences in Donbas have been exposed to since the conflict began. Media has, therefore, contributed to increased distrust, confusion and uncertainty for people in Donbas, in addition to the vulnerabilities and systematic risks that emerge in conflict contexts.

This chapter examines how amidst the conflict and ongoing humanitarian crisis communities in Donbas have been navigating different media, accessing news sources and making sense of the diverse and often contradictory information that is presented to them through the media.[1] A particular focus for the analysis is placed on local media and their use by communities in Donbas against a backdrop of deteriorating trust in national media.

The Conflict in the Donbas Region

Armed tensions in the Donbas region began in April 2014 with the seizure of government buildings in Donetsk and Luhansk oblasts by armed pro-Russian activists (BBC, 2014). These actions followed a series of dramatic developments in Ukraine that were galvanized by the Euromaidan protests—a popular movement against the then-President Viktor Yanukovych's decision to halt negotiations with the European Union over an Association Agreement. During these protests violence by police towards protesters enhanced support for the protest movement, eventually leading to the overthrow of Yanukovych's pro-Russian government in late February 2014. A new government was then formed that immediately put forward a pro-European agenda. From the very beginning, the Euromaidan protests were criticized by the Russian government (Guardian, 2014; Miazhevich, 2014; RFERL, 2013). Following the ousting of President Yanukovych, the Russian government took the decision to annex Crimea under the pretext of defending the rights of ethnic Russians in Crimea from the new Ukrainian government.

[1] The chapter is based on the data obtained during the study 'Media Consumption and Assessment of Social and Political Processes in Ukraine by the Residents of Eastern Regions (Government-Controlled Areas of Donetsk and Luhansk Oblasts)', which was commissioned by Detector Media, an NGO.

At the same time, tensions escalated in the eastern regions of Ukraine where pro-Russian activists, allegedly supported by the Russian state, declared the sovereignty of the so-called Donetsk and Luhansk 'People's Republics' (known as DNR/LNR in Ukraine) and also held 'self-rule' referenda on their independence. While observers pointed out that developments in Donetsk and Luhansk seemed to be repeating the Crimean situation, the events progressed differently. Russia refrained from hastily incorporating the newly proclaimed DNR/LNR into the Federation as it had been speculated would occur, and following an unrecognized 'referenda', the Ukrainian government launched a military operation in the Donbas region (DW, 2014).

These tensions eventually led to a full-fledged war that by 2017 had claimed the lives of around 10,000 people and injured more than 23,500 (UNHCR, 2017). The conflict has also led to the displacement of 1.5 million people (UNHCR, 2018). The fiercest fighting between the Ukrainian army and separatist forces, which have received not only weapons and ammunition from Russia, but also occasional direct military support from regular Russian troops, occurred in the summer of 2014 and winter of 2015. Periods of intense fighting on the frontline were followed by the implementation of the Minsk I and II ceasefire agreements that required the withdrawal of heavy weapons and provided a roadmap for the provisional settlement of the conflict. While these agreements have largely contained the fighting, sporadic clashes have continued. Parts of Donetsk and Luhansk oblasts remain under the control of the self-proclaimed DNR/LNR and are cut off from the rest of the region by a 427-kilometre-long 'contact line'.

At the time of writing, the conflict is now in its fifth year and is accompanied by a dire humanitarian crisis that is affecting around five million people in Eastern Ukraine. Hundreds of thousands of Donbas residents, especially those living in remote communities, suffer from interrupted water and electricity supplies (UN, 2018). According to a needs assessment study conducted by Caritas Ukraine, the deficit of drinking water is significant issue for people living in Donbas (Caritas Ukraine, 2019). A number of communities do not have access to a centralized water supply. Moreover, unexploded ordinance constitutes another risk for almost two million people, according to UN estimates (UN, 2018). The conflict is also having an impact on economic activities and living standards, with unemployment, for example, higher in Donbas than in other parts of Ukraine (UN OCHA, 2019b). Older people are a particularly vulnerable group and make up 30% of people in need in the conflict-affected region (UN OCHA, 2019b), with many now lacking access to basic healthcare.

Debates Over the Origins and Nature of the Conflict

Numerous attempts have been made to explain the roots, causes and triggers for the conflict from various perspectives including political, media and academic domains, the latter of specific interest for this chapter. Some scholars point to the decisive role of external aggression, while others focused on other factors, namely, the historical roots of the conflict, the role of identity, economic pressures, the peculiarities of political culture in the Donbas region and the contribution of local elites (e.g., Giuliano, 2015, 2018; Kudelia, 2017; Matveeva, 2016; Wilson, 2016). Identity cleavages and populism have provided a central focus for analysis. Researchers with an interest in Ukraine have long explored and debated regional diversity in the country and its impact on political processes (e.g., Kulyk, 2011; O'Loughlin, 2001; Osipian & Osipian, 2012; Riabchuk, 2015; Sasse, 2010; Shulman, 2005). The Donbas region has often been at the centre of such discussions and analysis for a number of reasons. As a densely populated region, it was one of the leading industrial centres in the Russian empire and Soviet Union (Wilson, 1995), which defined its strategic importance. According to the State Committee of Statistics of Ukraine (2018), about 6 million people still reside in Donetsk and Luhansk oblasts, despite recent population declines due to the conflict, and this still accounts for 14% of the total population of Ukraine. As a borderland region, Donbas has been populated by a significant number of ethnic Russians. Around 30% of those living in Donbas identify as ethnic Russian and 11% identify as both Russian and Ukrainian (Giuliano, 2018). At the same time, Donbas residents are commonly believed to also gravitate towards a 'soviet-type' of identity in cultural terms, rather than a 'Russian' one (Hrytsak, 1998; Wilson, 2016).

The collapse of the USSR precipitated a severe crisis in this industrial region. The 1990s were marked by ubiquitous poverty, increased levels of crime and the emergence of a new 'business-industrial elite'—a group of people who capitalized on the privatization of previously state-owned assets. The trajectory of post-Soviet transformation in Donbas has shaped its political culture. Kudelia (2017), for instance, identifies a weak civil society, political inertia and a strong quest for paternalism among its defining features.

Such peculiarities of regional identity and political culture are manifested in certain political and geopolitical views and preferences, as well as voting patterns. For instance, Donbas has long leant towards closer ties with Russia rather than the West and has largely opposed applications for NATO membership. It is also notable that such allegiances have been instrumentalized by political forces, especially those originating from the region, such as the Party of Regions and its leader Viktor Yanukovych, which have emphasized Donbas's exclusivity and opposed the 'nationalist policies' of political opponents (Kudelia, 2017). This has contributed to a stronger regional identity in Donbas.

Distorted coverage of the Euromaidan protests in both Russian and some Ukrainian media reinforced pre-existing divisions and helped to mobilize support for separatism. Additional support arguably stemmed from broader local concerns, including the economic consequences of broken ties with Russia and 'a sense of abandonment by Kyiv', in the words of Giuliano (2018, p. 158). Yet, amongst Donbas residents, support for separatism was low. According to survey data, less than a third of respondents surveyed in Donetsk and Luhansk backed separation from Ukraine (Giuliano, 2018). Under such conditions, some scholars have argued that local elites tightly connected to Yanukovych and Russia's intervention played an important role in contributing to the outbreak of conflict (Wilson, 2016), with local police and security services under their control failing to contain the violence or in some cases joining the separatist forces. Wilson (2016) also points out that the Yanukovych 'Family' provided key financial resources for the separatist movement.

The role of mass media in contributing to the conflict has also been much debated, and scholars have identified the discursive contestation and conflicting narratives that have emerged through media coverage of the conflict (e.g., Baysha, 2018; Cottiero, Kucharski, Olimpieva, & Orttung, 2015; Ojala & Pantti, 2017; Pasitselska, 2017; Szostek, 2018). In this context, it is plausible then that the vulnerability of communities in the Donbas region has been aggravated by exposure to media content that is highly manipulative and has presented conflicting narratives.

Donbas Residents, Their Media Consumption and the Local Media Landscape

Debates over the role of disinformation and propaganda in instigating conflict in the Eastern Ukraine have prompted an increased interest in the study of media consumption among Ukrainians, especially in the Donbas region. A number of international and local non-governmental organizations have commissioned surveys to examine various aspects of media consumption (e.g., Detector Media, 2015a, 2015b, 2017, 2018; IMI, 2017a; U-Media, 2017). However, there is a lack of data prior to 2014 and the outbreak of the conflict, which complicates analysis of the media consumption dynamics.

While the existing data is mostly general, it helps us to understand the context around media use by Donbas residents and identify patterns in the structure of media consumption. Some of those follow national trends, such as declining TV consumption, with 88% of respondents watching TV on a regular basis in 2012 and 73% in 2015, and an increase in online news consumption, up from 41% in 2012 to 57% in 2015, in Donetsk oblast (U-Media, 2015).

Consumption of online media has seen a steady growth in Donbas in recent years. Strikingly though, residents of Donetsk oblast have preferred local online news media to national media, which differs from other regions of the country, where the use of local online websites increased from 51% in 2012 to 69% in 2015 (U-Media, 2015), 28% higher than other regions of Ukraine (Detector Media, 2016). Conversely, internet users in Donetsk oblast are the least interested in national news websites in comparison to respondents from other regions. These trends illustrate the increased significance of local news media for Donbas residents after the escalation of conflict in 2014–2015. At the same time, they also suggest a continuing detachment from the national media discourse and a greater interest in local issues.

Another major difference in the media consumption habits of Donbas residents concerns Russian media. According to one of the studies, 71% of respondents from Donetsk oblast reported watching Russian TV news in 2014, which was the highest proportion among all the regions in Ukraine (Detector Media, 2016). However, a number of other surveys have confirmed that consumption of Russian media has declined since 2014 (Detector Media, 2015b, 2016), to just 8% and 9% of residents in Donetsk and Luhansk oblasts, respectively, reporting consuming Russian media in 2018. Such declines could be explained by both the implementation of restrictive measures towards Russian TV channels by the Ukrainian government and lower levels of trust in the Russian media (U-Media, 2015). Survey data support this by showing lower trust in Russian TV channels among residents of Donetsk oblast, down from 57% in 2014 to 9% in 2015.

A crisis of trust has extended to Ukrainian media as well. Although trust in Ukrainian national and regional media has been significantly higher, many polls record a general decline in trust in the media after 2014. The last three years, however, have shown some positive signs, with trust in national TV channels increasing from 36% in 2015 to 53% in 2018 in Donetsk oblast and from 27% to 33% in Luhansk oblast. Despite this, the evidence shows persisting uncertainty, anxiety and mistrust shape public attitudes in the region. Insights from the qualitative study that illustrate these characteristics are presented later in the chapter.

This review of survey data helps to sketch a general picture of media consumption in the conflict-affected region but does miss significant nuances. One concerns the diversity in terms of media consumption, both between Donetsk and Luhansk oblasts and within each oblast. For instance, one of the studies found that 51% of respondents in Donetsk reported consuming local television in early 2017 (IMI, 2017b), twice as many as in Luhansk oblast (IMI, 2017c). The same study identified considerable differences within each oblast. For example, only 23% of respondents from Volnovakha reported watching local TV, compared to 53% in Mariupol, a city with well-developed local media. A number of other examples illustrate the diversity of experiences amongst local communities in the Donbas region, differences which can be attributed to the local media landscapes, broader structural conditions, such as economic development, their proximity to the frontline and also their own experiences of the conflict.

Despite a lack of research on the media landscape of the Donbas region, analyses by media NGOs in Ukraine have identified several important trends. Namely, media in the region has been marked by uneven levels of market development and instrumentalization of media by the owners with close links to local political elites. In such conditions, local media have tended to act as mouthpieces for their owners rather than serve the interests of their communities. Conflict and occupation of the capital cities of both oblasts have also led to the closure of local media outlets, further weakening the region's media mix.

News Consumption and Perception in Conflict-Affected Donbas: Insights from the Qualitative Study

The previous sections outlined the context of news consumption among residents of the Donbas region following the escalation of armed conflict in 2014. This section presents the findings from an empirical study that explored how residents of Donetsk and Luhansk oblasts select their news sources and interpret information obtained from the media, especially with regard to the conflict that has unfolded in their region. Qualitative by the nature of its research questions, the study aimed to examine different experiences and views of Donbas residents with regard to news consumption, interpretation and perception. The data for analysis was obtained from 9 focus group discussions and in-depth interviews with 48 families that were conducted in 8 different locations in government-controlled parts of Donetsk and Luhansk oblasts during the summer of 2017. In total, 173 people participated in the study. They were selected to represent different age groups, gender, types of employment and residence.

The analysis revealed a number of patterns in participants' attitudes to the media and news content. It identified a pervasive scepticism towards the media and the news and information they provide, as well as adopting a position of distrust by default. 'You can't trust anyone these days' emerged as a common theme from the discussions and interviews. At the same time, participants of the study were inclined to regard interpersonal communication as a source of 'trustworthy' information.

> *Well, I probably trust not some kind of media or something else, but I rather trust my friends who are street-smart, so to say (...) I trust their judgements.* (22-year old female respondent, Pokrovsk, Donetsk oblast)

Interestingly, media personalities and bloggers were largely viewed as more credible sources compared to media organizations. Some respondents also identified local opinion leaders as reliable sources of information.

Participants explained that their scepticism and distrust towards the mainstream media was due to observing inaccuracies in the content, especially in the coverage of the conflict, and a perception that the media serves the interests of political elites and their owners, rather than people. In explaining their distrust of media, participants occasionally noted a gap between media coverage and their own experiences of the conflict, which suggested a lack of awareness of their problems by the authorities, media and, presumably, also citizens from other regions. Such perceptions have contributed to a sense of alienation from the media against a backdrop of broader alienation from the state. Comments from participants also revealed that many of them perceive mainstream media to be part of an establishment, one that they are largely opposed to.

In addition to frustration, participants' responses showed a low level of deliberation about news source selection. Few participants could rationalize their media choices. In a similar vein, the vast majority of participants could not explain how they reached a conclusion about the trustworthiness of information and credibility of media. One explanation offered by a 60-year old male respondent from Pokrovsk town in Donetsk oblast is illustrative in this regard.

> *Well, I trust... It's like in life, for what reason do we trust people? Well, I feel like trusting, so I trust...*

The analysis suggests that judgements on media credibility and, hence, information trustworthiness are often guided by intuition and shaped by existing beliefs and opinions. Intuitive loyalty was often implied by participants that tended to trust Russian media.

> *They lie on Russian [TV channels] too. Both these [Ukrainian TV channels] and those, but those [Russian TV channels] are more truthful.* (Female respondent, Hirske, Luhansk oblast)

Alienation and frustration with the mainstream media have also been driven by the conflicting narratives, which different media have presented.

> *TV channels just inflame people against each other... I watch one TV channel and they show things one way; someone watches another TV channel—and they show everything some other way... People are being pumped up with such (contradictory) information and then people don't understand each other. We can conclude that such people are easy to manage.* (Male respondent, Zolote, Luhansk oblast)

The study also found that while some participants were generally supportive of Ukrainian or Russian narratives regarding the conflict, the majority had rather ambivalent attitudes to both mainstream narratives, against a backdrop of general distrust, fatigue and accumulated disenchantment. Participants within this group tended to voice a 'middle-ground' narrative that intertwined some elements of mainstream narratives with other ideas. In fact, this 'middle-ground' narrative may be described as an umbrella term for various claims, emphasis, and varying degrees of confidence or uncertainty and so on. One of the crucial commonalities for different versions within the 'middle-ground' narrative concerned the recognition of Russia's interference and its geopolitical interest in Ukraine. Yet, discontent was largely directed towards the Ukrainian government. Within this 'middle-ground' narrative, the conflict is widely perceived as 'imposed' by the powerful, with ordinary people caught up in someone else's game.

Accordingly, many participants tended to perceive mainstream media as an instrument of elites. It is noteworthy that despite widely reported frustrations with the mainstream media, the majority of study participants continued consuming their content. Few citizens turned to local media as reliable news providers. Based on participants' reported attitudes and experiences, it seems plausible to conclude that local media were unable to provide a credible alternative to the national media. This could be explained by several factors. Firstly, many local media do not provide the news and information that audiences are seeking. As a consequence, their content often does not satisfy the needs and interests of local communities. One focus-group participant from the village of Zolote in Luhansk oblast said, "Well, people buy *Popasnyanskiy vestnik* (translated as 'Popasna newsletter', a local newspaper), but mostly because of the TV guide", which suggests that the guide is the most interesting content found in the paper. Participants repeatedly noted that information found in the local media, especially the local press, is outdated and dull. Secondly, the use of local media by local elites and administrations has also engendered a critical attitude towards these providers. A number of participants explained how local media simply act as a mouthpiece for local government and politicians, undermining their credibility.

Thus, although participants were generally familiar with the local media, that is, they recalled the names of providers and reported occasionally consuming their content, local media did not stand amongst the broader media landscape, which frustrated respondents. The failure of local media to provide relevant, timely information and low trust and credibility as a news provider encouraged a number of participants, especially younger respondents, to turn to social media instead. In particular, to the local community pages on the social network site VKontakte, a Russian platform that is very popular in the post-Soviet countries was cited by participants as their primary source for local news and useful information.

> *Definitely, when I need information about things happening in our city—I usually look online, on Vkonktakte. There are many groups related to Mariupol there. For instance, there's a group "Mariupol. A Roll Call". And when there are some bangs, blasts… or something is happening, I definitely go to this page and try to get the information.* (Female respondent, Mariupol, Donetsk oblast)

Notably, such groups and community pages are popular not only in large towns, but also in smaller towns which sometimes do not have any local news outlets. For these communities, such groups are important media sources.

The study thus found that local groups and community pages on social media performed some of the functions of local media and therefore were able to meet certain needs of local communities in the Donbas region, predominantly timely and up-to-date information, which is crucial for conflict-affected or disaster communities. Such local group and community pages are also able to provide a space that fosters a sense of community, which is crucial for Donbas residents, given the frustration and alienation felt by conflict-affected communities, alongside the broader socio-economic challenges. At the same time, local groups and community pages on social media only have a limited reach, especially amongst those communities with poorly developed infrastructure and those that are unable to access these platforms, such as older and impoverished people. The analysis of the Donbas case shows that the absence of a credible local media may contribute to frustration, resentment and disconnectedness for conflict-affected communities. Moreover, the vulnerabilities for conflict-affected or disaster communities could potentially be mitigated by local media better serving the needs of such communities. To what extent dynamically emerging forms of local media, including local groups on social media can provide an alternative to traditional local media deserves a separate in-depth exploration.

References

Baysha, O. (2018). Synecdoche that kills: How Barack Obama and Vladimir Putin constructed different Ukraines for different ends. *International Communication Gazette, 80*(3), 230–249.

BBC. (2014, April 7). Ukraine crisis: Protesters declare Donetsk 'republic'. *BBC*. Retrieved November 25, 2018, from https://www.bbc.com/news/world-europe-26919928.

Caritas Ukraine. (2019). Needs assessment of the water, sanitation and hygiene supply in buffer zone of Donetsk and Luhansk oblasts of Ukraine. Retrieved April 5, 2019, from https://www.humanitarianresponse.info/sites/www.humanitarianresponse.info/files/documents/files/caritas_ukraine_needs_assessment_2019.pdf.

Cottiero, C., Kucharski, K., Olimpieva, E., & Orttung, R. W. (2015). War of words: The impact of Russian state television on the Russian Internet. *Nationalities Papers, 43*(4), 533–555.

Detector Media. (2015a, March 20). What is the level of support of the messages of Russian propaganda in the conflict area: Sociological survey. *Detector Media*. Retrieved November 20, 2018, from https://ms.detector.media/mediaprosvita/research/yakim_e_riven_pidtrimki_mesedzhiv_rosiyskoi_propagandi_u_zoni_konfliktu_sotsiologichne_opituvannya/.
Detector Media. (2015b, July 1). Ukrainian mass media lose trust, but attitude to Russian media declined critically, even in the East. Sociological survey. *Detector Media*. Retrieved November 25, 2018, from https://ms.detector.media/mediaprosvita/research/ukrainski_zmi_vtrachayut_doviru_ale_stavlennya_do_rosiyskikh_zmi_kritichno_pogirshilos_navit_na_skhodi_sotsopituvannya/.
Detector Media. (2016, March 9). Media consumption and trust to media in the Donbas: Internews study. *Detector Media*. Retrieved November 26, 2018, from https://detector.media/infospace/article/113397/2016-03-09-mediaspozhivannya-ta-dovira-do-zmi-na-donbasi-doslidzhennya-internews/.
Detector Media. (2017, February 13). Survey of Russian propaganda influence on public opinion in Ukraine findings. *Detector Media*. Retrieved November 26, 2018, from https://ms.detector.media/detector_media_en/reports_eng/survey_of_russian_propaganda_influence_on_public_opinion_in_ukraine_findings/.
Detector Media. (2018). Opposition to Russian propaganda and media literacy: Results of all-Ukrainian opinion poll. *Detector Media*. Retrieved November 24, 2018, from https://detector.media/doc/images/news/archive/2016/136017/DM_KMIS_engl__WEB-2.pdf.
DW. (2014, May 2). Ukraine: 'Military operation begins in eastern city of Slovyansk'. *Deutsche Welle*. Retrieved November 27, 2018, from https://www.dw.com/en/ukraine-military-operation-begins-in-eastern-city-of-slovyansk/a-17606340.
Giuliano, E. (2015). The social bases of support for self-determination in East Ukraine. *Ethnopolitics, 14*(5), 513–522.
Giuliano, E. (2018). Who supported separatism in Donbas? Ethnicity and popular opinion at the start of the Ukraine crisis. *Post-Soviet Affairs, 34*(2–3), 158–178.
Guardian. (2014, March 17). Russian propaganda over Crimea and the Ukraine: How does it work? *The Guardian*. Retrieved November 28, 2018, from https://www.theguardian.com/world/2014/mar/17/crimea-crisis-russia-propaganda-media.
Hrytsak, Y. (1998). National identities in post-Soviet Ukraine: The case of Lviv and Donetsk. *Harvard Ukrainian Studies, 22*, 263–281.
IMI. (2017a). Analytical report on the results of sociological survey of residents of Southern and Eastern oblasts of Ukraine, conducted by the GfK in 2016–2017. Institute of Mass Information. Retrieved November 19, 2018, from https://imi.org.ua/wp-content/uploads/2017/07/IMI-sociology-final-report.pdf.
IMI. (2017b). Media preferences and political views in Donetsk oblast. Institute of Mass Information. Retrieved November 19, 2018, from

https://imi.org.ua/wp-content/uploads/2017/07/Media-Poll_w2_DonetskaNGCA_UKR.pdf.

IMI. (2017c). Media preferences and political views in Luhansk oblast. Institute of Mass Information. Retrieved November 19, 2018, from https://imi.org.ua/wp-content/uploads/2017/07/Media-Poll_w2_Luhanska_UKR.pdf.

Kudelia, S. (2017). The Donbas rift. *Russian Social Science Review 58* (2–3), 212–234.

Kulyk, V. (2011). Language identity, linguistic diversity and political cleavages: Evidence from Ukraine. *Nations and Nationalism, 17*(3), 627–648.

Matveeva, A. (2016). No Moscow stooges: Identity polarization and guerrilla movements in Donbass. *Southeast European and Black Sea Studies, 16*(1), 25–50.

Miazhevich, G. (2014). Russia Today's coverage of Euromaidan. *Russian Journal of Communication, 6*(2), 186–191.

O'Loughlin, J. (2001). The regional factor in contemporary Ukrainian politics: Scale, place, space, or bogus effect? *Post-Soviet Geography and Economics, 42*(1), 1–33.

Ojala, M., & Pantti, M. (2017). Naturalising the new cold war: The geopolitics of framing the Ukrainian conflict in four European newspapers. *Global Media and Communication, 13*(1), 41–56.

Osipian, A. L., & Osipian, A. L. (2012). Regional diversity and divided memories in Ukraine: Contested past as electoral resource, 2004–2010. *East European Politics & Societies, 26*(3), 616–642.

Pasitselska, O. (2017). Ukrainian crisis through the lens of Russian media: Construction of ideological discourse. *Discourse & Communication, 11*(6), 591–609.

RFERL. (2013, December 9). Ukraine's 'Euromaidan' through the lens of Russian television. *Radio Free Europe/Radio Liberty*. Retrieved November 19, 2018, from https://www.rferl.org/a/ukraine-euromaidan-russian-media/25194912.html.

Riabchuk, M. (2015). 'Two Ukraines' reconsidered: The end of Ukrainian ambivalence? *Studies in Ethnicity and Nationalism, 15*(1), 138–156.

Sasse, G. (2010). The role of regionalism. *Journal of Democracy, 21*(3), 99–106.

Shulman, S. (2005). National identity and public support for political and economic reform in Ukraine. *Slavic Review, 64*(1), 59–87.

State Committee of Statistics. (2018). The number of population (according to estimates), as of October 1, 2018. Retrieved November 26, 2018, from http://ukrstat.gov.ua/operativ/operativ2018/ds/kn/kn_u/kn0918_u.html.

Szostek, J. (2018). Nothing is true? The credibility of news and conflicting narratives during "Information War" in Ukraine. *The International Journal of Press/Politics, 23*(1), 116–135.

U-Media. (2015). Survey of media consumers: Donetsk oblast. Retrieved November 23, 2018, from https://www.slideshare.net/umedia/ss-53370234?from_action=save.

U-Media. (2017). Media consumption survey in Ukraine 2017. Retrieved November 24, 2018, from https://internews.org/sites/default/files/2017-09/USAID_UMedia_AnnualMediaConsumptionSurvey_2017_FULL_eng.pdf.

U-Media. (2018). Media consumption survey in Ukraine 2018. Retrieved November 23, 2018, from https://internews.in.ua/wp-content/uploads/2018/09/2018-MediaConsumSurvey_eng_FIN.pdf.

UN. (2018). Four years of conflict in Ukraine leave 4.4 million people in a dire humanitarian situation. *The United Nations*. Retrieved April 5, 2019, from http://www.un.org.ua/en/information-centre/news/4326-four-years-of-conflict-in-ukraine-leave-4-4-million-people-in-a-dire-humanitarian-situation.

UNHCR. (2017). Report on the human rights situation in Ukraine 16 February to 15 May 2017. Office of the United Nations High Commissioner for Human Rights. Retrieved November 25, 2018, from https://www.ohchr.org/Documents/Countries/UA/UAReport18th_EN.pdf.

UNHCR. (2018). Registration of internal displacement. Office of the United Nations High Commissioner for Human Rights. Retrieved November 19, 2018, from https://app.powerbi.com/view?r=eyJrIjoiZDU3MDEyNDgtZWNhNy00ODBmLWI0ZDYtMmJiZDQyMzExZGJkIiwidCI6ImU1YzM3OTgxLTY2NjQtNDEzNC04YTBjLTY1NDNkMmFmODBiZSIsImMiOjh9.

UN OCHA. (2019a). Ukraine Humanitarian Fund—Factsheet (March 2019). United Nations Office for the Coordination of Humanitarian Affairs. Retrieved April 5, 2019, from https://www.humanitarianresponse.info/en/operations/ukraine/document/ukraine-humanitarian-fund-%E2%80%93-factsheet-march-2019.

UN OCHA. (2019b). Ukraine: 2019 humanitarian needs overview summary. United Nations Office for the Coordination of Humanitarian Affairs. Retrieved April 5, 2019, from https://www.humanitarianresponse.info/sites/www.humanitarianresponse.info/files/documents/files/ukraine_2019_hno_summary_en.pdf.

Wilson, A. (1995). The Donbas between Ukraine and Russia: The use of history in political disputes. *Journal of Contemporary History, 30*(2), 265–289.

Wilson, A. (2016). The Donbas in 2014: Explaining civil conflict perhaps, but not civil war. *Europe-Asia Studies, 68*(4), 631–652.

CHAPTER 9

"Bloodbath, Invasion, Massacre": *Idoma Voice* and the Framing of the Farmer-Herder Conflict in Benue State, Nigeria

Confidence Uwazuruike

The farmer-herder clashes in Nigeria, that are of particular concern in the north-central region (also called the Middlebelt), have received increased attention from the media in the last five years (Duru, 2018; Ndubuisi, 2018). In Agatu, 300 people were reported dead following a herder-farmer clash on February 24, 2016 (Abonu, 2016). The same report claimed that more than 4000 lives had been lost in the north-central region of the country due to the farmer-herder conflict. A report by the International Crisis Group (ICG) published in July 2018 cited the violence between Nigerian herders and farmers as the cause of more than 1300 deaths between January and June 2018; this is six times the number of civilians killed by Boko Haram in the same period. It is also estimated that in the same six months period, 300,000 people fled their homes due to the conflict (ICG, 2018). In addition to its direct impacts, the conflict has led to further degradation of arable land and the loss of livestock upon which communities rely.

C. Uwazuruike (✉)
Preston, UK

J. Matthews, E. Thorsen (eds.), *Media, Journalism and Disaster Communities*, https://doi.org/10.1007/978-3-030-33712-4_9

Like most conflicts in Nigeria, the conflict between farmers and herdsmen has sometimes been seen as having ethnic and religious undertones. It is easy to see why this might be inferred. The herders involved in the attack are mostly from the Fulani ethnic group and are therefore often referred to as *Fulani Herdsmen*, referencing their primary ethnicity and distinguishing them from farmers who are non-Fulani. The *Fulani Herdsmen* are also predominantly Muslim as opposed to the farmers who are mainly Christian, at least in the north-central region of Nigeria. The Nigerian Government's perceived inadequate response to the conflict has been attributed to this ethnic consideration, with the Nigerian President Muhammadu Buhari both a Fulani Muslim and a herder himself. Thus, the ethnoreligious narrative is sometimes tied to the political one. Importantly, this is a view shared by many from the communities affected by the violence. For instance, Higazi (2016, p. 370) avers that "Plateau indigenes claim that there is an Islamic agenda to dominate Plateau State and that Muslims instigated violence on the Jos Plateau in their struggle for power".

Perspectives such as these, however, ignore that previous relationships and dealings between Muslims and Christians in the Middlebelt were mainly those of coexistence and not violent conflict (Higazi, 2016). Fulani and northern leaders also reject the equation of Fulani with herdsmen or the notion that the conflict is some sort of Islamic conquest (Osaghae, 2017). The President of Nigeria himself disputes the description of the perpetrators of this violence as herdsmen, explaining that herdsmen would not arm themselves with AK-47s but instead the sticks and machetes that they use to cut foliage for their herds (Sahara Reporters, 2018). He argues that the real perpetrators are not herders but insurgents from the Libya conflict. For the communities affected by the violence, however, herdsmen are seen as Fulani (Osaghae, 2017).

Conflicts between the Fulani nomads and those living on and working the agricultural land in Nigeria are not new. Evidence suggests that such disputes were recorded as far back as the 1800s (Lott & Hart, 1977). The recent conflict has, however, resulted in more deaths than all previous incidences combined and is occurring in communities that have seldom experienced conflict in the past. The escalation of tensions between farmers and herdsmen has been attributed to the impacts of prolonged conflict and terrorism in northern Nigeria but also a consequence of the intersection of other drivers of insecurity, namely, climate change, and specifically desertification, population growth, and increased demands for food production (Abbass, 2012; Bello, 2013; Maiangwa, 2017; Osaghae, 2017). These vulnerabilities combine to limit land that is available for farming and for grazing livestock, a dispute which is at the centre of the ongoing tensions between herders and farmers. Marietu and Olarewaju (2009) argue that it is land disputes, primarily between herdsmen and farmers over grazing lands, which is the primary source of conflict that contributes to insecurity for communities in the Benue Valley.

This chapter explores how this conflict is covered by local news media in Nigeria. It focuses on reporting in the *Idoma Voice*, a local newspaper that is available in Benue State in north-central Nigeria, one of the areas disrupted by the farmer-herder conflict. Three major ethnic groups live in Benue—Idoma, Tiv, and Igede. The most high-profile attacks by herdsmen, however, have occurred in Idoma land, including the Agatu killings of 2016. The focus for the newspaper is to report primarily on news and events in Idoma land and Benue but it also covers national news.

Studies examining news media coverage of conflict in Nigeria have tended to focus on the national press (e.g. Ette, 2012; Uwazuruike, 2018b; Yusha'u, 2015). This study, however, aims to offer insights into conflict reporting by a newspaper produced by and for a community directly affected by violence in Nigeria.

Framing Conflicts

Media scholars have consistently stated that the media play an important role in conflict, not limited to the passive reporting of events as they happen but extending to the active planning, waging, legitimising, understanding, historicising, and execution of any conflict (Hoskins & O'Loughlin, 2015). Thus, the shift from mediation to mediatisation. The news values inherent in conflict—drama, violence, negativity, impact, audience identification, ability to register on the public agenda—create a symbiotic relationship between the media and conflict, such that relevant actors in conflicts cannot ignore the media and the media cannot ignore conflicts. The greater media attention in Nigeria in recent years on the farmer-herder conflict, which followed an increase in the violence, is illustrative of this symbiotic relationship.

The nature of conflict makes objective reporting almost impossible (Carruthers, 2011). Patriotism, propaganda, national interest, and ethnic solidarity are all elements of conflicts that make objectivity impossible, especially in an ethnically diverse country like Nigeria (Uwazuruike,

2018b). As Nyamnjoh (2015) argues, African journalists practise "patriotism to the home village" when reporting national conflicts. Several studies of the Nigerian news media show how news organisations mostly reflect ethnic and regional interests when reporting on conflict. Yusha'u's (2015) study shows that Nigerian newspapers followed regional lines when reporting the crises in the Niger Delta. Similarly, Uwazuruike's (2018b) study of Nigerian press coverage of the Chibok schoolgirls' abduction reveals a similar level of bias.

One way to understand situation-specific reporting of conflicts is through studying how conflicts are framed or presented to audiences by news media. The news media can, through selection and emphasis, moderate how the audience perceives a news event (Entman, 1993; Gamson, 2005; Reese, 2010). Entman (1993) states that frames select aspects of a perceived reality and make them more salient in a text such that a particular problem definition, causal interpretation, moral evaluation, and/or treatment recommendation is promoted. Thus, a frame does not just provide meaning; it draws boundaries, sets up categories, and defines what is in and what is out in an active process (Reese, 2007).

This study uses framing theory to examine how *Idoma Voice*, as a local newspaper in Benue State, frames the farmer-herder conflict in north-central Nigeria. While there are four integral locations in the process of framing—the communicator, the text, the receiver, and the culture (Entman, 1993; de Vreese, 2005)—it is recognised that journalists play a significant role in the news framing process. They oversee the process of selection of the observed reality, by choosing what events and issues to write about and sources to interview and then organise material into a comprehensible story format, sometimes drawing on the frame repository in a given society, such as common understandings of a conflict, its history, and progression that is held by their audience (Van Gorp, Vettehen, & Beentjes, 2009; Brüggemann, 2014).

Methodology

To identify common frames in the coverage of the farmer-herder conflict, this study applies a qualitative critical analysis to a small sample of content gathered from the *Idoma Voice*. The *Idoma Voice* newspaper is a local newspaper which provides news for the Idoma ethnic group, one of the three dominant ethnic groups in the present-day Benue State in Nigeria. The print version of the newspaper was first published in July 2009, with an online version launched in 2013. It is published in English and read mainly by residents of Benue State.

News stories were extracted from the online version of *Idoma Voice* newspaper (idomavoice.com) using the advanced option of Google search. The search string used was *herdsmen* and searches were conducted for articles published between 2013 and 2018. The stories retrieved from 2013 to 2018 were screened for relevance and produced a final sample of 44 news stories for analysis, as shown in Table 9.1. The small sample size in this five-year period might be due to the newspaper's primary emphasis on print and its slow expansion into the online space. This study has, however, not analysed coverage in its print editions to ascertain if there is similar low level of coverage as the online editions. Interviews and opinion pieces were not included in the sample as the research was interested in ascertaining the news frames adopted by the newspaper.

Extracted news stories were read by the researcher to determine news frames. The unit of analysis was a full news story including headlines, pictures, and the main body of the text. The critical analysis followed three interconnected steps. First, news articles were carefully read and the central topic, narrative structure, recurring themes, and framing devices were noted. News articles were then revisited to note for reasoning devices—roots, consequences, appeals to principle, and remedies. The type of sources and their usage in news texts was also noted. These were then interpreted by considering relevant contexts, cultural norms, and political events to generate frames. When more than one news frame is present in a story, the main frame is referred to as the primary frame and the alternative frame is referred to as the secondary frame. Frames tended to be prominent in the headlines and initial paragraph.

Table 9.1 News reports on farmer-herder conflict extracted from idomavoice.com

Year	*Number of news reports*
2013	1
2014	8
2015	5
2016	12
2017	5
2018	13

Framing the Farmer-Herder Conflict

The analysis of news stories reporting the herdsmen-farmers conflict in the *Idoma Voice* reveals three dominant but overlapping news frames: a framing of the conflict as *bloody violence and mass death*, a frame of *disruption to normality* and a cover for criminal activities, and a framing of the conflict as requiring *government intervention* (Table 9.2).

From the data, the most common news frame is the conflict as *bloody violence and mass deaths*. More than half of all analysed news stories adopted this frame (*n* = 25, 56%). *Disruptions*, which were usually a consequence of this violence, was the next most common (*n* = 15, 33%) news frame. The one frame which does not deal directly with violent acts was the *government intervention* frame which was found in 11% (*n* = 5) of stories. The framing of the conflict in *Idoma Voice*, thus, concentrates more on the violent aspects of the conflict and its associated consequences, especially deaths. News stories, therefore, tended to be mainly episodic, limited to the event and not written to consider broader issues such as the underlying causes of the conflict.

Framing the Conflict as Bloody Violence and Mass Death

News stories containing this frame presented the conflict as mass and endless killings mainly perpetrated by the herdsmen on the host community (e.g. "Bloodbath: Day Fulani Herdsmen turned Benue community to slaughter field, 55 feared killed", *Idoma Voice*, 25 March 2014; "BLOOD, BLOOD EVERYWHERE! Another 63 killed in Benue by Fulani

Table 9.2 Distribution of frames used in the coverage of the farmer-herder conflict in *Idoma Voice*

	Bloody violence	*Disruption*	*Government*	*Total*
2013	1	0	0	1
2014	4	2	1	7
2015	3	2	0	5
2016	8	2	1	11
2017	3	3	0	6
2018	6	6	3	15
Total	25	15	5	45
	56%	33%	11%	100%

Herdsmen", *Idoma Voice*, 26 March 2014). Stories of reprisals and mass burials and condemnation of violence were also illustrative of this frame.

Words used to describe *the violence* were usually provocative, such as *bloodbath, slaughter, carnage, bombard, gruesome, invasion, senseless, massacre*. In the articles, perpetrators were, sometimes, referred to as [hired or suspected] *mercenaries, gunmen, and invaders*. Most times, the violence was presented as unprovoked. For instance, in the 24 March 2014 story: "BLOOD, BLOOD EVERYWHERE! Another 63 killed in Benue by Fulani Herdsmen", the following eyewitness quote demonstrates:

> The painful part of this attack is that communities in Gwer have had no prior incidents with the herders.

All headlines that adopted this frame presented the Fulani herdsmen as being responsible for the violence. Sometimes, however, the main body of reports was not as conclusive. For example, a story published on 7 May 2013 used the headline: "Fulani Herdsmen invade Agatu, kill five, sack 8 villages," thus implying that Fulani herdsmen were the aggressors. A closer reading of the main body of the story, however, shows that the herdsmen acted in retaliation following the death of an 11-year-old boy and as their livestock was stolen.

> VC gathered from eyewitnesses that crisis broke out in Agatu after some yet to be identified persons allegedly killed 11-year-old Abdulkair Mohammed and forcefully took away 10 cows from a Fulani herdsman. (*Idoma Voice*, 7 May 2013)

In the same story, the Miyetti Allah Cattle Breeders Association of Nigeria (MACBAN) suggested that the herdsmen were simply trying to retrieve their cattle rather than invading the village. Another discrepancy occurred in a story published on 25 March 2014 ("Bloodbath: Day Fulani Herdsmen turned Benue community to slaughter field, 55 feared killed", *Idoma Voice*) where words such as *bloodbath* and *slaughter* were used in the headline, with the violence attributed to Fulani herdsmen. The main body of the story, however, was less sombre, referring to the perpetrators as "suspected mercenaries".

Thus, there is a sense that the reports were tilted to convey a certain understanding of events especially in their headlines. A report from 16 May 2016 further emphasises this. The headline, "We killed Benue youths because they stole our cows—cattle breeders", presents cattle breeders as valuing their livestock over human lives. The main body of the story, however, cites the herders' association condemning violence in the host community and stating their ignorance of the particular attack.

> This is highly regrettable because they should have notified the security apparatus already put in place by the state government who could have handled the matter. They didn't report the matter even to us. (MACBAN)

The quote shows that the herder association encouraged the perpetrators to report matters to the police instead of engaging in reprisal attacks on youths who were alleged to have stolen "800 cows belonging to the resident herders". This represents a different attitude amongst the herders towards the violence and loss of life than is suggested by the headline. The demonisation of herders, especially in the headlines of news stories, is consistent with what Wolfsfeld, Frosh, and Awabdy (2008) describe as the victim mode of conflict reporting, where journalists think of their own people as victims and demonise the perceived aggressors. The problem, however, is that demonisation of the other may only serve to encourage further violence (especially reprisals from the victim community) and makes alternative less violent means of resolutions more difficult to reach (Uwazuruike, 2018b).

Another feature of this frame is that, even when deaths could not be solely attributed to the herdsmen, they were still often blamed for the deaths. In one story, in which unknown gunmen attacked a number of villages in Benue ("Benue bleeds as gunmen kill 20 in Zaki Biam", *Idoma Voice*, 20 May 2017), the following statement was added:

> The incident is coming barely one week after suspected Fulani Herdsmen killed over 10 people in Buruku area of the state, which led to the governor, Samuel Ortom issuing them 72 hours ultimatum to vacate the area.

Such statements at the very least suggest a connection to the violence by Fulani herdsmen. Stories of community reprisals were also attributed to Fulani herdsmen. An example of this was found in an article published on 17 May 2017: "Fulani herdsmen invasion: Four killed as Igede community boils". Here, the headline does not suggest that the story may have been another reprisal following the deaths of three Fulani men. The phrase "Fulani herdsmen invasion" paints the herders as the aggressors.

Framing the Conflict as Disruption and Crime

Articles that depicted a shift from the normal way of life were illustrative of this frame. They include those identifying the negative consequences arising from the violence, calls for self-defence and community violence, crimes associated with violent activities of herdsmen such as sexual violence, robbery, arson, devastation, and possible terrorism.

This frame described the violence perpetrated by the herders as beyond agitation for grazing rights or reprisals for cattle rustling. In a story published on 22 May 2014, "Fulani Herdsmen rape 2 women in Benue", herders were presented as "invading the houses of the victims when their husbands were out and took turns to rape the women". They "also made away with N60,000 as well as two handsets belonging to the women". News stories like this presented the herdsmen as criminals and cast shadows on the claims that the conflict was about grazing rights. A similar story concerning sexual violence ("'He promised to give me N700 after sex'—14-year old girl narrates after being raped by herdsman", *Idoma Voice*, 16 May 2018) was presented as an "abomination" on the land that might make the soil infertile. The community leader who espoused this view was given more prominence in the news story than the victim, and this served to amplify the crime as a desecration of the entire community. The story also described herdsmen as destroying farms, restricting the activities of most villages who were farmers, and threatening the peaceful coexistence of the community.

Other reports, such as one that reported how a community leader was killed when experimenting with local charms ("35-year-old man killed while testing gun charm in Owukpa", *Idoma Voice*, 21 July 2014), emphasise how the community had been "at war" with Fulani herdsmen "over the latter's refusal to leave their land". Other reports of disruption show how herdsmen's violent activities lead to homelessness and hunger ("Idoma people rendered homeless by Fulani Herdsmen dying of hunger", *Idoma Voice*, 18 March 2015).

> Yesterday, scores of displaced persons were seen with their surviving family members, personal belongings and domestic animals, trekking to some neighbouring communities in Apa LGA of the state.

This resulted in an increase in internally displaced people (IDP) and suspicion and discord amongst communities (e.g. "Agatu Sole Administration, Mike Inalegwu under fire for claiming Ortom forced him to sell land to

Fulani Herdsmen", *Idoma Voice*, 21 January 2017). Other news stories that exemplified this disruption were contained in calls for self-defence, linked to perceived government indifference ("Killings in Nigeria: Dr. Paul Enenche of Dunamis speaks again, calls for self-defence", *Idoma Voice*, 13 January 2018).

Government Reaction

This frame is evidenced by stories that chronicled the government's reaction to the conflict and the actions of the police towards the violence. This was the least common frame and demonstrates, broadly, that newspaper coverage tends to focus more on the violence and its impacts on communities.

Stories adopting this frame generally showed the government, including state governments, as adopting a neutral position towards the conflict, especially in articles published in the earlier years of the conflict, for example, by reporting on government efforts to diffuse tensions by establishing committees involving the farmers and herders to help reach solutions and resolve disputes ("Boko Haram cannot attack schools in Benue—Suswam", *Idoma Voice*, 28 May 2014). In one story, the governor of Benue was keen to stress this collaboration between farmers and herders:

> I've set up a very powerful committee made up of herdsmen and farmers. The Fulanis and Benue farmers are working together. Where there is any incidence, the committee intervenes. For instance, it was the Fulanis that helped in arresting the people that raped a woman. We believe we will sustain the peace.

The quote above also shows the positive action by Fulanis in seeking to prevent violence and also government efforts to promote peace.

Most stories that illustrate this frame, especially when a government representative did not feature as a source in the report, was critical of the government and how it was failing in its responsibilities to safeguard communities. Thus, the *government reaction* frame sometimes occurred as a secondary frame within stories that were categorised *with bloody violence* as the primary frame (e.g. "Again, Fulani herdsmen attack Agatu, kill over 50", 1 March 2016; "Fani Kayode, other Nigerians applaud Dr. Paul Enenche for speaking against killers in Nigeria", *Idoma Voice*, 18 January 2018). Criticisms were reserved mainly for the federal government, due to their responsibilities for ensuring security. A political decision taken by the state governor of Benue to join another political party distinct from the party in power at the federal government was presented in news reports as a necessary step to address the farmer-herder conflict. Therefore, the state government itself was attributing rising insecurity to the lack of an appropriate policy response by the federal government.

> Leading the decampees, Dura said they defected to the PDP to join forces with Governor Samuel Ortom to stop the planned annihilation and take over of the Benue Valley. (10,000 APC members dump party for PDP in Benue, *Idoma Voice*, 2 September 2018)

This was a departure from the initial position taken by state governments towards the conflict, which was to emphasise peaceful solutions through collaboration. During this period state governments enacted anti-grazing bills which criminalised open grazing, which only served to increase the discord between herders and farmers (Nasir, 2018). Such decisions, at the very least, demonstrated a lack of coordination between federal and state governments on how to handle the conflict.

The *Idoma Voice* was also critical of the state government's role. One specific concern was over the safety and welfare of IDPs.

> About 48 hours after the heinous attack, the Benue State government has not opted to set up refugee camps and provide relief materials to the displaced persons. This is even as Governor Gabriel Suswam and his deputy, Steven Lawani, are yet to visit the destroyed areas to assess the level of damage and console the affected families. ("Idoma people rendered homeless by Fulani Herdsmen dying of hunger", *Idoma Voice*, 18 March 2015)

Therefore, while criticism of the federal government centred on its failure to provide security and respond to the violent attacks, the state government was denounced for its lack of support for the communities affected by the violence.

Further Discussion and Conclusion

The analysis of coverage of the farmer-herder conflict in *Idoma Voice* shows that the newspaper primarily reported on and emphasised the violent aspects of the conflict. The news value of violence is, undoubtedly, one reason violence emerged as a prominent frame. Moreover, for the *Idoma Voice*, as a local community newspaper whose primary readership does not include decision-making elites, it is understandable that it focuses on the effects of violence due to its impact on and relevance to its readership. The aim of adopting this editorial line, therefore, may be to inform their audience about the deadly realities of the situation they face, galvanise them to defend their communities, or raise grievances and protest against government policies concerning the conflict. In focusing on the violence, however, the *Idoma Voice* is consistent with the notion of conflict as a news value, which is often sensationalised to improve circulation (Lee & Maslog, 2005). Proponents of peace journalism, as an alternative to traditional modes of conflict reporting, argue that the media can be used to encourage peaceful conflict settlement and serve as mediators of peacebuilding and reconciliation processes (Galtung, 2003; Kempf, 2007). The analysis here indicates that the *Idoma Voice* was unable to fulfil this role.

The framing of news reports in *Idoma Voice* also suggests that the *Idoma Voice* preferred the victim mode of reporting (Wolfsfeld et al., 2008), which is primarily evident through the demonisation of the Fulani herdsmen. Reporting violent acts perpetrated by the group is an effective way of demonising the group. This is also consistent with conflict reporting practices in other parts of the continent where journalists practise "patriotism to the home village" (Nyamnjoh, 2015, p. 36) and support their own side in a conflict. The findings, however, show that frames that demonise the other do not promote alternative strategies, especially those that may encourage non-violent solutions.

Moreover, as reports were focused on the immediate violence around the conflict and their consequences, news frames tended to be more episodic than thematic, failing to explore the causes of the conflict or provide in-depth analysis beyond the proximate realities, such as information about the number of deaths. Relevant discourses around open grazing, ranching, climate change, and land laws and acquisition, which are the antecedent conditions for conflict between farmers and herders in Nigeria, were seldom reported. Reports in *Idoma Voice* that emphasised violence without considering solutions to the conflict, its history, or context, arguably, played a part in the reactions of state governments during the latter period analysed in the study—with the state government enacting anti-grazing bills and suggesting that the federal government was complicit in exacerbating the conflict. As Hoskins and O'Loughlin (2015) argue, the mediatisation of conflict can influence the actions of actors in conflict situations. The possible effects of mediatisation enhance the case for a peace journalism model that is adopted by local or community media especially when reporting on conflicts that impact on their audience.

There is a paucity of personal or human-interest stories in the *Idoma Voice*'s coverage. This is similar to findings in studies that analysed the coverage of conflicts in the Nigerian national press (e.g. Uwazuruike, 2018a, 2018b). Like the national press, news stories in the *Idoma Voice* simply report on matters of process, declaring how many people were killed in an attack without personalising these stories, for instance, by naming victims. The similarity here with the national press is at odds with practices in other journalism and cultural contexts where stories are personalised, for example, in the Israeli press where reports include a "roll-call of the dead" (Frosh & Wolfsfeld, 2007, p. 113). Although, as a local newspaper, *Idoma Voice* might not try to encourage empathy to distant audiences, as members of the community already experience the conflict, humanising the victims of the violence dignifies their deaths and helps other members of this disaster community to empathise with the families of victims. By treating victims as merely statistics, the newspaper's reporting fails to humanise victims and foster community solidarity. This finding confounds expectations as given its proximity to the people, a local newspaper like *Idoma Voice* would be expected to tell the personalised stories of the victims from their community that have lost their lives in the violence.

Finally, there was no evidence to show that the *Idoma Voice* drew on ethnoreligious frames. Considering the actors in this conflict—northern Fulani Muslims and southern Benue Christians—the ethnoreligious frame was one that would be expected to occur in news coverage. Studies of the Boko Haram insurgency in the national press in Nigeria, for instance, show that this is a convenient frame for journalists to adopt (Uwazuruike, 2018a, 2018b). The problem with ethnic frames, as argued in those studies, is that they are deeply polarising and simplistic. The lack of ethnoreligious frames in *Idoma Voice* is an interesting finding, demonstrating that local journalists that are familiar with issues within their own community might not resort to the simplistic ethnic frames that are often found in the national or international press. This is an area to be investigated in subsequent studies.

References

Abbass, I. M. (2012). No retreat, no surrender: Conflict for survival between Fulani pastoralists and framers in Northern Nigeria. *European Scientific Journal, 8*(1), 331–346.

Abonu, A. (2016). The Agatu Massacre. *This Day (4/April/2016) online edition* [online]. Retrieved April 8, 2019, from https://www.thisdaylive.com/index.php/2016/04/04/the-agatu-massacre/

Bello, A. U. (2013). Herdsmen and farmers conflicts in North-Eastern Nigeria: Causes, repercussions and resolutions. *Academic Journal of Interdisciplinary Studies, 2*(5), 129–139.

Brüggemann, M. (2014). Between frame setting and frame sending: How journalists contribute to news frames. *Communication Theory, 24*(1), 61–82.

Carruthers, S. (2011). *The media at war* (2nd ed.). Basingstoke: Palgrave Macmillan.

de Vreese, C. H. (2005). News framing: Theory and typology. *Information Design Journal, 13*(1), 51–62.

Duru, P. (2018, July 15). O Benue, it is 500 dead in 196 days! *Vanguard*. https://www.vanguardngr.com/2018/07/o-benue-it-is-500-dead-in-196-days/

Entman, R. M. (1993). Framing: Toward clarification of a fractured paradigm. *Journal of Communication, 43*(4), 51–58.

Ette, M. (2012). 'Nigeria as a country of interest in terrorism': Newspaper framing of Farouk Abdulmutallab, the underwear bomber. *Journal of African Media Studies, 4*(1), 45–59.

Frosh, P., & Wolfsfeld, G. (2007). ImagiNation: News discourse, nationhood and civil society. *Media, Culture & Society, 29*(1), 105–129.

Galtung, J. (2003). Peace journalism. *Media Asia, 30*(3), 177–180.

Gamson, W. A. (2005). Projections of power: Framing news, public opinion, and U.S. foreign policy. *Public Opinion Quarterly, 69*(2), 324–326.

Higazi, A. (2016). Farmer-pastoralist conflicts on the Jos Plateau, central Nigeria: Security responses of local vigilantes and the Nigerian state. *Conflict, Security and Development, 16*(4), 365–385.

Hoskins, A., & O'Loughlin, B. (2015). Arrested war: The third phase of mediatization. *Information, Communication & Society, 18*(11), 1320–1338.

ICG. (2018). *Stopping Nigeria's spiralling farmer-herder violence*. Report 262/Africa. Retrieved April 7, 2019, from https://www.crisisgroup.org/africa/west-africa/nigeria/262-stopping-nigerias-spiralling-farmer-herder-violence

Kempf, W. (2007). Peace journalism: A tightrope walk between advocacy journalism and constructive conflict coverage. *Conflict & Communication Online, 6*(2), 1–9.

Lee, S. T., & Maslog, C. C. (2005). War or peace journalism? Asian newspaper coverage of conflicts. *Journal of Communication, 55*(2), 311–329.

Lott, D. F., & Hart, B. L. (1977). Aggressive domination of cattle by Fulani Herdsmen and its relation to aggression in Fulani culture and personality. *Ethos, 5*(2), 174–186.

Maiangwa, B. (2017). "Conflicting Indigeneity" and farmer–Herder conflicts in postcolonial Africa. *Peace Review, 29*(3), 282–288.

Marietu, T., & Olarewaju, I. (2009). Resource conflict among farmers and Fulani herdsmen: Implications for resource sustainability. *African Journal of Political Science and International Relations*, 3(9), 360–364.

Nasir, J. (2018, May 19). Miyetti Allah vows to resist Benue anti-grazing law, asks Buhari to warn Ortom. *The Cable*. Retrieved from https://www.thecable.ng/miyetti-allah-vow-to-resist-benue-anti-grazing-law-asks-buhari-to-caution-ortom

Ndubuisi, C. I. (2018). HTS Teologiese Studies/Theological Studies. Retrieved April 8, 2019, from http://www.hts.org.za

Nyamnjoh, F. (2015). Media and belonging in Africa: Reflections on exclusionary articulation of racial and ethnic identities in Cameroon and South Africa. In W. Mano (Ed.), *Racism, ethnicity and the media in Africa* (pp. 28–55). London: I.B Tauris.

Osaghae, E. (2017). Conflicts without borders: Fulani herdsmen and deadly ethnic riots in Nigeria. In P. Aall & C. Crocker (Eds.), *The fabric of peace in Africa: Looking beyond the state* (p. 49). McGill-Queen's Press.

Reese, S. (2010). Finding frames in a web of culture: The case of the war on terror. In P. D'Angelo & J. Kuypers (Eds.), *Doing news framing analysis: Empirical, theoretical and normative perspectives* (pp. 17–42). New York: Routledge.

Reese, S. D. (2007). The framing project: A bridging model for media research revisited. *Journal of Communication, 57*(1), 148–154.

Sahara Reporters. (2018, April 30). Buhari: Herdsmen don't carry AK-47. *Sahara Reporters*. Retrieved from http://saharareporters.com/2018/04/30/buhari-herdsmen-don't-carry-ak-47

Uwazuruike, C. (2018a). Reporting Boko Haram: Framing the Chibok schoolgirls' abduction in the Nigerian press. *African Journalism Studies, 39*(3), 66–84.

Uwazuruike, C. (2018b). Reporting terrorism Boko Haram in the Nigerian press. Bournemouth University. Retrieved April 24, 2019, from http://eprints.bournemouth.ac.uk/31179/

Van Gorp, B., Vettehen, P. H., & Beentjes, J. W. J. (2009). Challenging the frame in the news: The role of issue involvement, attitude, and competing frames. *Journal of Media Psychology, 21*(4), 161–170.

Wolfsfeld, G., Frosh, P., & Awabdy, M. T. (2008). Covering death in conflicts: Coverage of the second intifada on Israeli and Palestinian television. *Journal of Peace Research, 45*(3), 401–417.

Yusha'u, M. J. (2015). Regionalism and ethnicity in the Nigerian press: An analysis of the coverage of Boko Haram and the Niger Delta conflicts in the Guardian and Daily Trust. In W. Mano (Ed.), *Racism, ethnicity and the media in Africa* (pp. 137–156). London: I.B Tauris.

CHAPTER 10

Media and Reconciliation: A Study of Media-Led Initiatives in Post-IS Mosul

Aida Al-Kaisy

In times of conflict and its immediate aftermath, the media's focus often tends to be on the battle itself, the victory of one side over another and the liberation of captured territories, rather than on the importance of reconstruction of physically and psychologically destroyed people and their communities (Cottle, 2012; Tumber & Prentoulis, 2003). When the self-proclaimed Islamic State (IS) was apparently driven out from the control of Mosul in July 2017, the Iraqi media followed a very similar pattern. National and international media coverage was dominated by jubilant scenes of the Iraqi armed forces liberating citizens of Mosul (e.g. see Chappell, 2017; Rudaw, 2017). In reality, the liberation of Mosul from IS in July 2017 has left the city with a considerable number of challenges and protracted physical, psychological and sociological reconstruction efforts. With the absence of any clear authority or political will at a national level to ensure that reconstruction includes reconciliation amongst communities, community-based and civil society-led initiatives to support reconstruction and reconciliation efforts have become widespread. These are

A. Al-Kaisy (✉)
School of Oriental and African Studies, University of London, London, UK

J. Matthews, E. Thorsen (eds.), *Media, Journalism and Disaster Communities*, https://doi.org/10.1007/978-3-030-33712-4_10

proving to be more effective in rebuilding the city's infrastructure, clearing mines as well as addressing other vulnerabilities that persist, such as attending to social and educational needs. Local media enterprises—mainly radio and online—have also begun to play a role much more significant than their national or even, international counterparts in reporting on the key issues that are impacting on socio-political progress in the newly liberated territories and for conflict-recovering communities.

This chapter examines the emergent forms of local media in post-IS Mosul, with particular emphasis on local radio in the governorate of Ninewah, in northern Iraq, where Mosul is the capital. The discussion is concerned primarily with the community of Mosul with some reference to the governorate of Ninewah, since part of the empirical work covered the governorate as well as Mosul itself. It will begin by considering the socio-political context in Mosul, the wider role that media have played in Iraq, particularly in Mosul and surrounding territories that were occupied by IS, examining media consumption habits and local media platforms. It offers a brief case study analysis of Radio Al Ghad, a community radio station launched in March 2015 that was able to provide crucial information and content during the occupation and examine their role in the current context for communities in Mosul rebuilding post-IS.

The chapter draws upon findings and insights from fieldwork conducted in Mosul and Erbil between March and April 2018 that formed part of a media development project, *Tasalah*, funded by the French Ministry for Europe and Foreign Affairs and jointly organized by Canal France International (CFI) and Media in Cooperation and Transition (MiCT). As part of this project, a series of events entitled #MosulTalks were organized. The aim of these events was to connect the various stakeholders working on reconciliation, reconstruction and reintegration efforts in the newly liberated territories of Ninewah in order to explore the role of media in supporting community cohesion and reconciliation in Mosul and Ninewah. The following chapter will explore the main themes of these events, which were education, transitional justice and reconciliation, in order to unpick what role local media are playing in providing spaces that might contribute to local and national governance and help to address the vulnerabilities for communities in post-conflict recovery.

#MosulTalks centred on two events: first, a series of workshops held in March 2018 at the Mosul Book forum, a library-cum-café that also serves as an events space for cultural and arts happenings, which brought together local CSOs, academics, media organizations and politicians from across Ninewah to consider the key issues in Mosul that relate to education, transitional justice and governance; second, a conference that was held in Erbil on 10 April, 2018 which opened up the discussion and debate about these challenges to the wider national and international community of organizations concerned with and working on reconstruction and reconciliation in Mosul. To update and further inform the analysis presented in this chapter, this fieldwork was supplemented with a number of interviews with key media stakeholders in Mosul and Ninewah province, which took place in December 2017 and January 2018.

The Socio-political Context in Mosul

When the Iraqi government proclaimed in December 2017 that it had driven IS out of Iraq, in it followed a long national and international offensive that contributed to the demolition of Mosul and its people. While figures vary wildly on the number of civilians killed in the violence perpetrated by both IS and coalition forces, with estimates of between 20 and 40 thousand deaths in Ninewah province (Iraq Body Count website, 2019; UNAMI, 2019), the city was almost completely destroyed. The city's infrastructure, housing and schools had inevitably been targeted by belligerents from all sides and there remain thousands of unexploded mines and devices left by IS.

The post-conflict situation in Mosul has been exacerbated the Iraqi government's minimal efforts to manage it and dwindling interest and concern from the international community, with the reconstruction of Mosul seen as a business project by many within that community

The issues of social reconstruction and reconciliation within communities in Ninewah province have been addressed with even less vigour. The social fabric of Ninewah has always been characterized by its ethnic and cultural richness: Christians, Kurds, Shabak, Shias, Sunnis, Turkmen and Yazidis were all driven out and persecuted under the IS occupation. The return of some of these groups to their former communities and what is left of their homes, as internally displaced peoples (IDPs) has highlighted the failure of national efforts to reintegrate and reconcile with those communities that remained under IS. Many of these groups have felt excluded from national dialogues and the processes of state formation and remain both misrepresented and under-represented in Iraq's politics and society (Haddad, 2011; Mansour, 2016). These issues are often cited as reasons for the emergence of radical doctrines, such as the ideology advanced by IS.

The fate of those who are considered to be IS collaborators and affiliates and the potential for their reintegration into society is also proving to be a huge challenge (Benraad, 2017). Alleged collaborators and their families continue to be subjected to violent retaliatory attacks in the absence of official mechanisms for transitional justice (Parry, 2018). At the #MosulTalks workshop on transitional justice held on 24 March 2018, the discussion around transitional, justice-related challenges in post-IS Ninewah highlighted two interlinked areas: retaliatory justice and justice for victims of IS. The question of how IS family members might be reintegrated into Ninewah was key, with the suggestion that detention camps for IS prisoners could act as ghettos and echo chambers for IS doctrine as well as have a long-term impact on the psychological development of children. At the same time, retaliation attacks on suspected IS members and their families, which began as far back as 2015, continue to be an uncontrolled issue, with a proliferation of hate speech across all media and digital platforms further exacerbating incitement and violence. During the workshop it was also noted that there was an absence of any clear legal framework to ensure legitimate accountability and prosecution for those responsible for IS crimes. This has been further aggravated by the sense of impunity for those have sought to take justice into their own hands.

While some international efforts are in place to rebuild and secure Ninewah, the question of responsibility looms as Iraq's political divisions and challenges remain. There are still a variety of different security forces allegedly providing safeguarding across the province. The international coalition, the Iraqi army and paramilitary groups such as Hashd Al Shabi or Popular Mobilisation Units are all vying for control of different areas and groups across Ninewah, which is contributing to insecurity (Al-Khafaji, 2019; Crisis Group, 2017).

The #MosulTalks workshop on governance, held in Mosul on the 25 March 2018, focused on a lack of transparency in official processes and institutions as well as the endemic corruption that exists at many levels, including local government. Participants noted that it was these very issues that contributed to the rise of IS in the province. Without adopting mechanisms to address them, long-term reconciliation and social cohesion efforts will fail to resonate. The lack of political will to develop solutions to combat underlying ethno-sectarian practices in government was emphasized as a major contributing factor to the deep lack of trust in public institutions and politicians.

Education has been another key challenge to reconstruction efforts in Ninewah. After capturing Mosul, IS seized control of the education system, closing a number of schools and replacing the Iraqi curriculum with its own more radical ideological syllabus. Universities were closed for long periods of time and tertiary students remained deprived of educational opportunities over the course of the occupation. Responding to IS attempts to propagate its radical ideology amongst Mosul's youths remains a huge challenge as children cope with PTSD and the psychological damage of radicalization and indoctrination. The need to safeguard them and all citizens in the province against the return of extremist ideologies is imperative.

In the workshops, participants also voiced serious concerns about the physical damage to schools and colleges. This has restricted access to many key educational facilities in the province, a factor which is now contributing to overcrowding in classrooms, particularly in neighbourhoods where people are returning post-IS. There also remains a shortage of academic resources for students, teachers and researchers, many of whom have not been paid regularly over the last few years. Poverty is having a direct impact on literacy, with illiteracy increasingly an issue for young women and girls. There is also an ongoing concern about how to manage the reintegration of IS-educated students returning to schools and colleges across the province. The continuing threat posed by IS, as well as a lack of security and economic opportunity, is fuelling a brain drain, with many educators leaving the province for other cities in Iraq and beyond. A key issue for academics that participated in the workshop was the absence of professional training programmes aimed at educational staff. They felt there was a lack of expertise and knowledge about how best to represent minority religions, sects and ethnicities in education establishments and curricula. It was also felt that there was an absence of any solutions-driven response from the Ministry of Education with regard to the reform of IS curricula.

In the main, these challenges have been addressed by loosely organized but, in some cases, vibrant civil society initiatives. They work in education, reconstruction, social cohesion and justice, filling the vacuum that a lack of clear governance structures has created. Within the course of rebuilding the city and reviving social norms, there have been some examples of local and community media in Mosul playing an important role as a source of information about reconstruction, public services and local government activities, as well as serving as a platform for exchange and dialogue amongst citizens, connecting them to politicians and civil society in the absence of other national platforms.

Media in Iraq

After the 2003 Iraq war and the overthrow of the Baathist regime, the transition government of Iraq issued 100 'Orders' or regulations that replaced all preceding Baathist political legislature and were meant to assist in building 'democratic institutions' in post-conflict Iraq. These orders lifted the ban on non-state authorized media, opening up the media landscape to an environment of pluralism and diversity. The expansion of Iraq's media landscape and facilitation of pluralism in the Iraqi media scene was however ill thought out. As an externally imposed process, it failed to consider Iraqi history and local context. After decades of state control, a loose sense of diversity and representation in non-state media outlets pervaded the landscape in Iraq. Private businessmen and politicians launched new television channels. A combination of internal and regional political and social factors, combined with a lack of proper regulation, led to the growth of a media environment that eventually served to exasperate the situation of conflict (Price, 2011). Al-Marashi (2007) describes this Iraqi public sphere, defined by pluralism and diversity of ownership, as 'ethno-sectarian media sphericules', aligning with Gitlin's (1998, p. 173) theory of 'public sphericules' that describes how media fragmentation and proliferation create multiple smaller public sphericules 'that enrich(es) the possibilities for a plurality of publics—for the development of distinct groups organized around affinity and interest'. His analysis of polarized ownership based on political and religious identities and affiliations offers some insight into the problems of media pluralism in Iraq.

According to a 2018 national audience qualitative survey conducted by BBC Media Action, Iraqi audiences are highly media literate; they are aware of media ownership, are sceptical of the media and consume their media accordingly. The study, conducted in July 2018 by BBC Media Action and shared with the author, also demonstrated that audience viewing habits have changed dramatically in the last few years. They now channel hop to find accurate news and engaging content. Social media are playing an even greater role in media consumption as audiences tend to remain loyal to programmes rather than to channels. Local media are also considered by many to be more accurate and informative.

Media in Mosul

The majority of media in Mosul were developed either under the occupation of the city by IS or in the aftermath. Those who broadcast and published during the occupation have reconsidered their content and outputs in the wake of reconstruction and reconciliation in the city. A further quantitative research study from BBC Media Action undertaken in 2019, which looked at audience consumption habits and perceptions of the media in Mosul since the departure of IS, has demonstrated the impact of local media on audiences in Mosul. Ninety-seven per cent of those surveyed by BBC Media Action have access to television and 79% claim to watch daily. There are two television channels aimed explicitly at citizens of Mosul. Al Mosuliya, now under the jurisdiction of IMN, is one of Iraq's most active regional television stations. It dates back to the times of Baathist rule and as such has a strong brand history and presence in Mosul. Ninewah Al Ghad is a newcomer to the Mosul media scene and is financed by Atheel al-Nujaifi, former Governor of Ninewah. Al Mosuliya has the highest audience figures with 85% daily viewers and 96% (484) weekly or more regular viewers.

After television, radio is the second most popular platform in Mosul; 32% of respondents in Mosul said that they have access to radio in their homes and 37% in cars. Fifteen per cent claimed to tune into radio daily, 40% weekly or more often, rising to 47% doing so at least once a month. The most popular radio station in Mosul is one that is local to the city: Radio Al Ghad that focuses its programming on local issues, culture and arts, successfully tapping into local 'Maslawi' heritage and identity. This case study is discussed further later in the chapter. Other radio stations that were referenced in the study included:

- *Al-Rasheed*: Owned by Saad Al-Janabi, a moderate Sunni politician with considerable wealth, broadcasts across Iraq, offering a mix of pan-national music and news content.
- *Radio Dijla*: One of the first commercial radio platforms to be founded in Iraq. Its programming consists mainly of music and talk show content with selected news bulletins. It is pan-national with good distribution, particularly in the western and northern provinces of Iraq.
- *Radio Nawa*: A national radio station, not specifically targeting Mosul, but with specialist programmes and content that are relevant

to Ninewah and surrounding areas occupied by IS. There were initially problems with their broadcast signal in some parts of Mosul, but this has improved since the liberation and Radio Nawa have reported accepting a growing number of participants in their live programming and call-ins from Mosul.

- *Radio Al-Mosul*: Part of the IMN, the national public broadcaster for Iraq. Much of its content focused on supporting the Iraqi army and counter-terrorism activities while Mosul was under occupation.
- *Start FM*: A station founded in 2017 with donor funding from Germany. It operates from Erbil in Iraqi Kurdistan and programmes are produced by a small team of radio professionals from Mosul. Its programming content focuses on Mosul and is aimed at a more liberal young audience, including 'educational and informational programmes for refugees from Mosul and the Ninewah province'.

Access to the internet is still more limited in Mosul compared to the rest of Iraq. While IS did not restrict access to the internet as such, they imposed high taxes on internet providers leaving the city with limited access to this day (See UN Habit Report, 2016). Despite this, a number of growing local online media initiatives are proving to be popular. Ein Al Mosul, or Mosul Eye (https://mosul-eye.org/), was a blog, written anonymously, that documented events in Mosul under the occupation of IS, providing citizens of Mosul and, perhaps even more so, diaspora and the international community with vital information and evidence of IS atrocities. It focuses now on the 'recovery' of Mosul, structurally as well as culturally. 'Mosul Eye was able to not only be a source of information but a social factor in the city', according to its founder Omar Mohamed (2018 cited in Guardian Podcast, 2018). After the battle, Mosul Eye played a different role, 'which is to rebuild civil society, trying to support the people who stayed in the city, trying to give them a voice, because they were voiceless'. A popular Facebook group—Bab al Toob—was also developed after the liberation of Mosul from IS. The page, with over 300k followers, shares photography and video footage from across Mosul that shows reconstruction in the city and music, art and cultural events. User-generated videos and content shared to the page that highlight example of Maslawi poetry, culture and heritage are extremely popular.

At the #Mosul Talks workshops, the role of the media in Mosul was a topic discussed at length by participants. The discussion focused on the challenges and indeed failure of the media to provide accurate and diverse information or indeed play a role in supporting the reconstruction efforts and social cohesion in Mosul today. It also identified a number of weaknesses that limits media's contribution.

These included recognizing that the media are vastly sectarian, partisan and politicized, with a lack of independent and impartial reporting across the board. Weak advertising markets and declining commercial revenues in the region see much of the private media owned by politicians and businessmen with particular agendas to promote, while state-financed public media acts mainly as a vehicle for government narratives. When IS first entered Mosul in June 2014, the government-funded *Al Iraqiya* focused its content on nationalistic images and songs to support the state and the army, rather than providing the public with the crucial information and access to sources that they needed to make informed choices about the conflict and their safety.

Discussants also acknowledged that much of the media's content is inaccurate, dominated by both misinformation and disinformation, which results in an absence of public trust in the media's handling of governance and politics. The number of casualties and deaths in battles or improvised explosive device (IED) attacks, for example, differs from platform to platform, leaving the public with very little understanding of critical events as they unfold.

Moreover, media are still not fulfilling their duties by holding officials to account or by tracking key issues related to governance and politics or by identifying corruption and mismanagement. Corruption in Iraq is considered to be an endemic problem and the media could play a very important role in tackling this (Transparency International, n.d.). The governor of Mosul since 2015, Nawfal Al-Akoub, has remained in place for years with little media coverage dedicated to exposing his corrupt practices, despite public outcry. It eventually took the sinking of a passenger ferry in Ninewah for parliament to finally dismiss him in March 2019.

It was also acknowledged that media are not serving their audiences by providing educational content. There are very few programmes aimed at supporting the learning of young people and children. There is also very minimal information aimed at supporting voters and citizens in order that they can fully contribute to and participate in elections and political processes.

One further weakness was a distinct absence of inclusive, balanced debate on issues of importance to Mosul and its citizens. The rise of the popular 'talk show' and the issues with partisan ownership has seen an increase in hate speech in the media. In Mosul, this tends to be ethnic, religious, political and even gender-based hate speech (Iraqi Media House, n.d.).

It was also suggested that the media are not lobbying government or civil society on behalf of the urgent requirements of the people of Mosul and Ninewah. As such, there is a dearth of diverse voices being heard and a lack of coverage on key issues, which ultimately reduces the media's role in supporting better governance and independent infrastructures. Attempts at building independent mechanisms that protect freedom of expression, those that emerge from civil society, have struggled to succeed without the support of the media.

A follow-up conference held in Erbil brought together media practitioners, academics and CSOs and NGOs working on reconstruction and reconciliation to develop a strategy that would enable the media to better support Mosul in its efforts to rebuild itself, physically and psychologically. A number of solutions were discussed. Whilst the focus has been on the physical rebuilding of Mosul, reconstruction should also consider the provision of services, the development of social cohesion as well as strengthening the socio-cultural aspects of society in Mosul. Participants discussed the significance of providing greater economic opportunities to minority and marginalized groups in Mosul. To this effect, it was suggested that the media could play a greater role in giving voice to minority groups across Mosul and beyond, providing an alternative to the prevailing narrative of the national media and looking at the challenges of reconstruction from a new perspective.

The question of how to support the reintegration of those who may have been formerly associated to IS, alleged IS collaborators, supporters and their families was also of concern. For some participants, there was doubt that reconciliation could ever be possible. For others, IS detention camps were not seen as a viable solution to peacebuilding. It was suggested that more work needed to be done to define what is meant by IS affiliation, taking a more grassroots approach, allowing local communities to be involved in setting the differentiations and meanings. It was confirmed by all participants that there was a lack of knowledge of the work that has been done by local and national CSOs in this field. It was suggested that stronger partnerships between the media and civil society would provide a fuller and more accurate picture of what is happening in this area. There is a lack of political will and motivation to support the rebuilding of Mosul and its links to the rest of the country. As such, public institutions and mechanisms were failing to deliver in areas such as justice which was resulting in violent acts of retribution with impunity. The media was called upon to play their role as the fourth estate, calling those in power to account for their actions and emphasizing issues of local concern. One practical suggestion was that media use their social media channels to highlight government failings and dearth of initiative. Hate speech is being disseminated by the mainstream media, and journalists need to develop an awareness of this and establish methods for identifying hate speech and avoid its propagation. It was also suggested that more support be provided to develop the capacity of local journalists in areas such as post-conflict reporting, social cohesion and peacebuilding. This needs to take place at both a professional and tertiary education level. Media literacy programmes were also seen as important in order to combat the rise of mis/disinformation. Enhancing trust in the media was considered to be integral to efforts towards rebuilding Mosul.

While the issues arising from the qualitative fieldwork of this study, as identified above, remain, there are examples where local media has played a more constructive role in the community. The following case study demonstrates how a local, community radio station is taking the initiative in providing a platform for inclusive and diverse dialogue for Mosul's residents.

Case Study: Radio Al Ghad

Radio Al Ghad was launched in March 2015 in response to the dearth of accurate information and reporting on Mosul for those still based in the city. Based in Erbil in the Kurdish Autonomous Region, just over 80 kilometres outside of Mosul, it began by broadcasting music at a time when IS had made internet and media access extremely difficult for the majority of Mosul's inhabitants. Despite repeated attempts by the militant group to jam the airwaves and stop their broadcasting, the station became known for its phone-in shows, allowing residents to interact and share current information about the occupation, warning people about areas where explosives and mines might be active and the hazards of everyday life in Mosul.

The station's, and in particular its phone-in shows', popularity is founded on the notions that citizens of Mosul have a very different perspective to that of national politicians and that their needs are unique compared to the rest of Iraq. This is a significant consideration as Mosul and the Ninewah governorate is home to one of Iraq's most diverse populations, many of whom have felt excluded from social and political life. In post-conflict Mosul, the unique needs of its citizens, as discussed earlier in this chapter, range from pragmatic needs in relation to reconstruction, education and justice, to those related to reintegration and reconciliation at a community level.

Phone-ins from the public have varied in their content under the occupation of Mosul by IS, but nearly all put their callers' lives at risk. Some callers would appeal for the Iraqi army and government to come and save them, assuring the government that they would support them in their battle. As fighting to recapture Mosul from IS became more intense in 2016, soldiers would call in to the station too, sending assurances to the people of Mosul that they would be safe. Al Ghad also took calls from the public, passing on word of civilian shelters to the armed forces in order that they did not become targets.

Al Ghad's manager, Mohamed Al Hashemi, suggested that radio was in fact the most effective media platform for them to choose in the case of Mosul, 'which is why Daish (IS) used Al Bayan Radio as their main means of communication inside Mosul' (Personal interview, 7 January 2019). Al Ghad's priority was to fill gaps in information and act as a public service by providing accurate news and countering rumours and disinformation spread by IS. During the war to liberate Mosul, their aims shifted as they identified a need to connect citizens with officials: 'We used to receive calls from people saying that they are waiting to be liberated and that they wanted the Iraqi forces to help them. Then we would get in touch with the Iraqi forces and use the station to send their messages to the people', Al Hashemi said. It was suggested that calls were more often made by women, in very hushed voices, who were under less scrutiny by IS.

Following the departure of IS from Mosul, they launched their now flagship programme, Sowt Al Muwatin or 'Voice of the Citizen', a live programme that has broadcast six days a week since 2017 and seeks to address the communication gap between government and the people on a more practical level. Al Hashemi reflects that without a media platform that reaches officials and citizens, extremists and radicals end up filling these spaces with their own information and gaining traction amongst those that are feeling marginalized:

> We raise problems on air and get solutions on air. Officials call in with their answers to people's questions. We have now developed a system to follow up

> on cases so we don't let any issues fall through the cracks, I do go and meet with municipality officials and follow up via an interview for example. Sometimes we can find an answer to an issue within six hours of the phone-in. It is improving citizen relations as they feel like they can have platform where they can speak. (Al-Hashemi, 2019)

In a society where the relationship between citizen and state is fractured, Al Ghad is offering a platform for political connectivity, where a citizen-state relationship can be facilitated.

Although privately funded from a number of different sources, a strong factor in its sustainability record, Radio Al Ghad follows the model of many community radio stations. It seeks to enable debate about local issues, including voices from the local community which tend to go unrepresented in the national media (Coyer, 2011). It acts as an information provider and, more importantly, as platform for cultural representation (Hall, 1997), a much-needed contribution to the Mosul reconstruction process. Al Hashemi was keen to stress that:

> We see Al Ghad as much bigger than a radio station when it comes to capacity and capability. We want to expand to other provinces but are also interested in start-ups and combining the media with economy making a direct connection with the private sector and local economy, working with people who have the intention to help youth and provide tangible solutions. (Personal interview, 2019)

Conclusion

There are echoes of what some scholars refer to as the development of an authentic public sphere with the local media in Mosul as they offer platforms to facilitate debate and discussion amongst citizens, although one has to be mindful of the Eurocentric assumptions about reason and political discourse that underpin the literature on the public sphere (Eickelman, 1999; Fraser, 1992; Negt & Kluge, 1993). Habermas's notion of the public sphere is a normative one: that there can exist a space, free of both state and market influence, where a free flow of information and ideas are transmitted through discourse and social interaction and this in turn allows for public opinion formation. However, in the case of this empirically based chapter, we see evidence of the potential development of public sphericules. Where the national media have failed to provide representation and relevant information for a society such as Mosul, one in dire need of both physical and psychological support in the aftermath of its occupation, the local media have taken some steps to fill those gaps. Al Ghad radio is one of a number of examples of civil society-led initiatives that are giving the people in Mosul a voice and platform for debate. During the conflict with IS, it acted as information provider as well as a space for interaction and debate, giving voice to those with little access to other platforms. In post-conflict Ninewah, it continues to offer citizens from across the spectrum an opportunity to call their government to account. The trust that was engendered during the conflict has been consolidated in a post-conflict context. Al Ghad's physical presence in Ninewah gives it the capacity to identify and thus answer the needs of local citizens.

As Iraq wrestles with debates about decentralization of government and services, so too do the media begin to reflect these similar narratives. The power of local media is evident in its consumption figures as local audiences turn to their community channels to satisfy information and representation needs. And yet, the role of the media in post-conflict processes such as reconciliation and reconstruction is often overlooked by political and social actors. In the case of Mosul and Ninewah, the needs of citizens are specific to their locality, ethnicity and religion. As citizens unite over issues related to services, local governance and the practicalities of reconstruction, there is a role for local media to play in providing platforms for information and debate around these issues.

References

Al-Hashemi, M. (2019). Interview by Aida Al-Kaisy, 7 January.

Al-Khafaji, H. (2019). Iraq's popular mobilisation forces: The possibilities for disarmament, demobilisation & reintegration. LSE Middle East Centre Report (November 2019). LSE Middle East Centre, London, UK.

Al-Marashi, I. (2007). The dynamics of Iraq's media: Ethno-sectarian violence, political Islam, public advocacy, and globalization. *Cardozo Arts & Entertainment Law Journal, 25*(1), 95–140.

BBC Media Action. (2019). The picture in Mosul—Reach and frequency of use of the media, attitudes and perceptions. Retrieved July 1, 2019, from http://dataportal.bbcmediaaction.org/site/.

Benraad, M. (2017). Mosul, Sunni Arabs and the day after. In A. Plebani (Ed.), *After Mosul Re-inventing Iraq*. ISPI. Retrieved July 1, 2019, from http://www.ledizioni.it/stag/wp-content/uploads/2017/10/Iraq_web_DEFDEF.pdf.

Chappell, B. (2017). Mosul has been liberated from ISIS control, Iraq's prime minister says. *NPR*. Retrieved July 1, 2019, from https://www.npr.org/sections/thetwo-way/2017/07/09/536307429/mosul-has-been-liberated-from-isis-control-iraqs-prime-minister-says; https://edition.cnn.com/2017/12/09/middleeast/iraq-isis-military-liberated/index.html?t=1561982489590.

Cottle, S. (2012). *Mediatized conflict: Understanding media and conflicts in the contemporary world*. Maidenhead: McGraw-Hill Education.

Coyer, K. (2011). Community media in a globalised world: The relevance and resilience of local radio. In R. Mansell & M. Raboy (Eds.), *The handbook of global media and communication policy* (pp. 166–179). Oxford: Wiley & Blackwell.

Crisis Group. (2017). Retrieved July 1, 2019, from https://www.crisisgroup.org/middle-east-north-africa/gulf-and-arabian-peninsula/iraq/188-iraqs-paramilitary-groups-challenge-rebuilding-functioning-state.

Eickelman, D. F. (1999). The coming transformation of the Muslim world. *Middle East Review of International Affair, 3*(3), 78–81.

Fraser, N. (1992). Rethinking the public sphere: A contribution to the critique of actually existing democracy. *Between Borders: Pedagogy and the Politics of Cultural Studies, 1*, 74–98.

Gitlin, T. (1998). Public sphere or public sphericules? In T. Liebes & J. Curran (Eds.), *Media, ritual and identity* (pp. 175–202). London: Routledge.

Guardian Podcast. (2018, September 26). The Mosul historian who risked his life to blog about life under Isis. Retrieved July 1, 2019, from https://www.theguardian.com/global-development/audio/2018/sep/26/mosul-historian-blog-life-under-isis-podcast-small-changes-omar-mohammed-mosul-eye.

Haddad, F. (2011). *Sectarianism in Iraq: Antagonistic visions of unity*. Oxford: Oxford University Press.

Hall, S. (1997). *Representation: Cultural representations and signifying practices*. London: Sage Publications and Open University.

Iraqi Body Count. (2019). Retrieved July 1, 2019, from https://www.iraqbodycount.org/database/.

Iraqi Media House. (n.d.). Retrieved July 1, 2019, from www.imh-org.com/.

Mansour, R. (2016). The Sunni predicament in Iraq. Carnegie Endowment for International Peace. Retrieved July 25, 2019, from https://carnegie-mec.org/2016/03/03/sunni-predicament-in-iraq-pub-62924.

Negt, O., & Kluge, A. (1993). *Public sphere and experience* (P. Labanyi, J. Owen Daniel, & A. Oksiloff, Trans.). Minneapolis, MN: University of Minnesota Press.

Parry, J. (2018). Legal pluralism and justice in Iraq after ISIL. POMEPS Studies 30 "The Politics of Post-Conflict Resolution". Retrieved July 25, 2019, from https://pomeps.org/legal-pluralism-and-justice-in-iraq-after-isil.

Price, M. E. (2011). Global media policy and crisis states. In R. Mansell & M. Raboy (Eds.), *The handbook of global media and communication policy* (pp. 180–191). Oxford: Wiley & Blackwell.

Rudaw. (2017). PM Abadi declares Mosul liberated from ISIS in victory speech. *Rudaw*. Retrieved July 1, 2019, from http://www.rudaw.net/english/middleeast/iraq/100720174.

The United National Assistance Mission in Iraq. (2019). *Facts and figures*. Retrieved July 1, 2019, from http://www.uniraq.org/index.php?option=com_k2&view=item&layout=item&id=945&Itemid=475&lang=en.

Transparency International. (n.d.). Iraq. Retrieved July 25, 2019, from https://www.transparency.org/country/IRQ.

Tumber, H., & Prentoulis, M. (2003). Journalists under fire: Subcultures, objectivity and emotional literacy. *War and the Media: Reporting Conflict, 24*(7), 215–230.

UN Habitat Report. (2016). Retrieved July 1, 2019, from https://reliefweb.int/sites/reliefweb.int/files/resources/UN-Habitat_MosulCityProfile_V5.pd.

PART III

Human (In)action and Humanitarian Crises

CHAPTER 11

Is Local Journalism Failing? Local Voices in the Aftermath of the Grenfell and Lakanal Fire Disasters

Kurt Barling

In the darkness of the early hours of June 14, 2017, shocking images of a tower block engulfed in flames flashed across our British television screens. Grenfell Tower, a 24-storey residential public housing block in the affluent London Borough of Kensington and Chelsea, had caught fire after an electrical fault in a refrigerator in one of the flats. The speed with which the fire broke out of the initial flat and travelled up the outside of the building overwhelmed the first responders from the London Fire Brigade (LFB) and bewildered bystanders. Whilst many residents survived by escaping the building, others became either trapped or followed long-standing advice issued by LFB and public authorities to *stay put* in the event of a fire. Distressing footage later emerged of tenants making face-time calls before perishing in the inferno. By sunrise the flames and billowing smoke were still visible across the London skyline and the smouldering hulk had claimed 72 victims.

K. Barling (✉)
Middlesex University, London, UK
e-mail: k.barling@mdx.ac.uk

J. Matthews, E. Thorsen (eds.), *Media, Journalism and Disaster Communities*, https://doi.org/10.1007/978-3-030-33712-4_11

In the hours and days immediately following the fire, that corner of West London was inundated with national and international news organisations wanting to understand what, why, how, where and when this disaster had happened. Above all, who and what was to blame for an entire tower block going up in flames in one of the world's richest cities. The post-event became a textbook example of journalism trying to get to the very bottom of the story. Virtually no politician, local or national, no media commentator, no public official responsible for building regulations, no public bystander was immune from the outrage that such a blaze could happen causing such a catastrophic loss of life.

The exhaustive media coverage including local outlets explored every imaginable issue that could have caused the fire, probing official decisions, public warnings from local residents' associations, the regulatory framework and the minutiae of construction decisions taken during the refurbishment of the block in the two years before the blaze. It was unfortunately investigative journalism that came too late for the victims. Journalism, it turns out, had seemingly failed to alert the residents, public and officials to the potential risks of a disaster in the making, when cladding, vulnerable to fire, was used to insulate the outside of the tower block.

On June 15, the day following the fire, British Prime Minister Theresa May quickly announced a public inquiry into the disaster, so blatant it seemed were the errors of the public authorities in preventing it from happening.[1] The disaster also raised very important questions for journalism and its practitioners. Why were these public safety failures not spotted beforehand? Was this due to the failure of grassroots journalism and a consequence of the economic failure of countless local newspapers? Was a deficit in local journalism putting the public at risk?

This chapter explores how local journalism failed to report on Grenfell before the blaze, ignoring the publicised concerns of the community, in particular those raised by the Grenfell resident's association about fire safety that followed a similar fire at Lakanal House in Camberwell, South London, in 2009. In this sense it will argue that Grenfell was a disaster foretold.

[1] On June 14, the author was among the first commentators to call for a public inquiry on BBC Radio 5 Live, BBC R4's PM Programme, BBC Radio London and the BBC News Channel.

What Do We Really Expect from 'Local Journalism'?

The journalistic response to the 2009 Lakanal House fire remains a cogent example of how local journalism can hold power to account. That disaster, where six people died, underlined the dangers of fire in tower blocks. It led to greater public awareness of the risks posed by fire to those living in high-rise residential buildings. This did not, however, stop the tragedy at Grenfell.

Holding power to account and keeping citizens informed about public affairs are integral to what is broadly characterized as the 'fourth estate' in power relations across liberal democracies. For more than a century an ecosystem of local journalism evolved which also fed the national storytelling appetites of the biggest selling newspapers and broadcasters. It has been credited with binding local communities together, shaping distinctive identities and, despite its imperfections, remained important and relevant to keeping readers and viewers alive to some interesting and important goings-on in their communities (Mair, Keeble, & Fowler, 2013; NAW, 2018; Nielsen, 2015). Local power relationships evolved in a way that fostered what you might call a *culture of exposure*. It was the fear of exposure of misdemeanours that provided local editors with real leverage.

At their best, "local newsbrands are a unique and powerful campaigning force which can bring about real change for the communities they serve" (NMA, 2018). The News Media Association, the trade body for local and regional newspapers in the UK, also claims that despite the public's growing scepticism about news providers more generally in the digital age, local papers remain the most trusted of all news sources, placed ahead of local commercial broadcasters, search engines, social media and other local websites. At the last 'census' there were still 1000 local newspapers in the UK (Local Media Works, 2019), and 65% of people who read a newspaper everyday read one of them.

Nevertheless, while editorial aspirations typically strove to scrutinise local town-hall decisions, cover crime and punishment in local magistrates and crown courts and expose local corruption, recent reality has fallen short of the democratic accountability of yesteryear. Despite aspirational values, journalism's declining potential has been widely critiqued by those observing the declining number of local news providers. *Guardian* commentator George Monbiot (2009) decried local journalism as one of the "most potent threats to British democracy, championing the overdog, misrepresenting democratic choices, defending business, the police and local elites from those who seek to challenge them".

In Britain, it is also increasingly the case that 80% of local media outlets are controlled by six large organisations including Newsquest and Trinity Mirror (Media Reform Coalition, 2019). Therefore, since these organisations work on economies of scale and often provide news from locations remote from the locality, it has become more difficult to represent the communities they serve.

The *Press Gazette* (2018), which has been charting local press decline over the past two decades, has reported on the closure of several hundred titles. In the wake of the Grenfell disaster, the intensity of the scrutiny of 'what went wrong' also fell on a failing local media. Emily Bell (2017) described the disaster as an example of "an accountability vacuum left by (a) crumbling local press". The assertion here is that greater scrutiny of the record of Kensington and Chelsea committee meetings and officials, in particular their response to active online groups, such as the Grenfell Action Group, a well-organised local community group that made representations to Kensington and Chelsea Council on behalf of local residents' about their concerns over refurbishment work, would have potentially placed pressure on the local authority to listen more attentively to its critics during the refurbishment of the tower blocks throughout 2015–2017. The warning signs, the argument runs, were missed by local news providers, specifically *Kensington and Chelsea News* (Preston, 2017a, 2017b; Feller, 2017).

It's clear that in the lead up to the Grenfell Tower blaze the online presence of the local residents' association and in particular the Grenfell Action Group blog were significant. However, a review undertaken by the author of local coverage shows that the local paper did not champion their cause and, as a result, other important outlets in the local news ecosystem failed to pick up on the story. It shows that there was very little local coverage of the issues being raised by the Grenfell Action Group before the fire. This is curious because local residents at Grenfell often used the disaster at Lakanal House to illustrate their concerns over the failure of the local housing authority to be transparent about, for example, their home's Fire Risk Assessment (FRA). Residents openly worried about the *probability* of repeating the mistakes at Lakanal and feared that only a disaster would bring a halt to the housing authorities' "negligence", as they saw it (Barling, 2017a).

What this discussion does remind us is that readers have grown to expect that local journalism demonstrates greater care for the community, understands the values of that community and prioritises solutions to their problems as much as identifying structural social, economic or political harms that may impact on their communities. At Grenfell, this may not have happened before the disaster. But, as I shall now argue, the reporting of the Lakanal House tragedy demonstrates that local journalism can provide strong public warnings about potential hazards and disaster risks even if they end up being ignored.

Reporting the Lakanal House Fire: Lessons and a Warning

Raphael Cervi was midway through his shift at the London Bridge restaurant on July 9, 2009, where he worked as a waiter, when he took the call from his wife Danielle. They were raising two young children in their home in one to the largest tower block estates in Camberwell. Their flat at the top of the block was well proportioned and spacious, with stunning views all the way to central London. Their neighbours were postmen, care workers and council officers and even fashion designers. Little did they know they were living in a death trap. Not because the building was structurally unsound, but rather because the policies and procedures designed to keep people safe in tower blocks were being regularly flouted (Barling, 2017b). It would take the fire disaster in 2009 to reveal how far the health and safety rot had penetrated.

Danielle was clearly in a panic. There was a fire in their block and she didn't know what to do. How could she get herself and the two children to safety? The local authority landlord had not made it clear to residents how they should behave in a fire. The basic guidance was to *stay put* and let firefighters get you out safely (Barling, 2017a).

By the time Raphael arrived back at Lakanal House, the LFB were already tackling the blaze, immediately declared a major incident. A *bridgehead*[2] had been established on the ninth floor, but had had to be evacuated as the fire spread below the firefighters. That was unprecedented

[2] This is how the forward firefighting station at a major incident is described. In a tower block this would usually be established within the building so that breathing apparatus and other essential firefighting equipment can be assembled close to the seat of the fire. At Lakanal, this proved impossible to maintain as the fire spread along the panels outside the building and also through cavities within it. As a result of the Lakanal blaze, firefighters training was revised to recognise that fires could not necessarily always be contained within the 'compartment' of the individual flat as originally envisaged, hence 'compartmentation'.

in tower block fires. Several flats on different floors had smoke billowing out of their windows. No firefighter could fathom why. Rafael rang his wife again, panic overwhelming him, for his own safety he was held back and was reassured that firefighters would be able to rescue them. They were wrong. He watched helplessly as his young family were consumed by the flames (Barling, 2017c).

By now chaos reigned in the operation to extinguish the fire, which was behaving unpredictably. It was moving sideways and downwards through the building. It was not being contained to the original flat where an electrical appliance had started the blaze. *Compartmentation* was supposed to give firefighters 60 minutes to reach and contain a flat fire. Over many years of legitimate refurbishments, this basic architectural fire-retarding feature had been compromised. Subsequently, this became a key feature of reporting in the local media for several years after the Lakanal fire (Barling, 2017). In fact, there is evidence that such local news coverage encouraged residents' associations across the UK, including the Grenfell Action Group, to ask more questions of their social landlords about fire safety in their own buildings. When in 2015 the major refurbishment project got underway at Grenfell, residents quoted Lakanal repeatedly. The Grenfell Action Group lobbied the local authority for access to information based on the recommendations of the 2013 Lakanal Inquest, which, unlike a public inquiry, was not able allowed to look into all factors that led up to the fire.

At Lakanal, firefighters got lost in the block in the tremendous heat and smoke of the blaze as it intensified. Worse still they had no known evacuation plans to hand. Each flat had been designed with a designated escape route along their external balcony to the main stairwell. Most residents had no idea that it was for that purpose; firefighters had no idea, nor did council officials when asked. Escape routes were often blocked. This became a matter the Lakanal inquest dwelt on in detail (Barling, 2017b). Although Grenfell Tower was a different design to Lakanal, the Grenfell Action Group remained concerned that there were also inadequate evacuation protocols in their block.

Local journalism revealed that Raphael's family perished needlessly in their flat because fire safety in tower blocks had been neglected, falling down the list of priorities for public and private landlords (BBC London News; *South London Press*; *London Evening Standard*; ITV London). It had been a disaster waiting to happen. Arguably it would have been impossible to have revealed all these dramatic oversights without the intensity of local newsgathering in the wake of the Lakanal disaster. These important insights were valuable to the residents' group at Grenfell but sadly were unable to prevent disaster.

At Lakanal House, the LFB could not initially explain the spread of the fire. Two fire experts, Sam Webb, an architect, and Arnold Tarling, a surveyor, appeared regularly in the local media suggesting that the external cladding and UPVC windows, which melt in extreme heat, may have contributed to the spread of the fire. Both experts had extensive experience in investigating the fire safety of buildings, yet they were regularly undermined in public by those whose interests they challenged. These experts turned out to be right, but during the initial phase of reporting on the Lakanal fire there were robust denials from the local authority, the company that installed the improvements and anyone else who had an interest to defend in the refurbished block. The details emerging early on prompted the local BBC newsgathering operation to invest resources in searching for answers (Barling, 2017c).[3]

BBC London was a very important part of the ecosystem of local newsgathering in the capital city and an important feeder of stories to their network news partners. Very quickly journalists identified the FRA document as an important source of safety information. Crucially, the FRA, as a legal requirement for landlords of multiple occupancy dwellings, should ensure fire safety by monitoring potentially risky structural changes to a building.

Newsgatherers at BBC London decided to issue a series of Freedom of Information requests, to identify when local authorities across London had carried out the last FRA in each of their tower blocks. It soon became clear to journalists that the LFB, the key fire regulator, had neither the FRA information nor did they know how many tower blocks were covered by their fire service across the capital. They also had no idea how many authorities were complying with the law by gathering this legally required information. Local authorities stalled, and some even refused to provide the information (Barling, 2009).

The BBC reported that the vast majority of local authorities had a very patchy idea of how compliant their tower blocks were with the Fire

[3] The author was the chief investigative journalist with BBC London at the time and ran a number of investigations into Lakanal over a period of four years. Longevity assists with credibility in getting sources to open up, but also brings with it a measure of hostility from those whom journalists investigate.

Regulatory Reform Order. This enabling legalisation, introduced in 2005, made all landlords (public and private) part of a self-regulatory regime for fire safety and removed the responsibility for inspecting premises from the LFB. According to a BBC investigation in 2009, it became clear that the system was poorly policed and even more poorly adhered to by public landlords (Barling, 2009). It also reported that it was inevitable that in such a lax regime people would eventually die again. The FRA issue flagged up at Lakanal became central to the Grenfell Action Group's search for answers from their local authority during the refurbishment of their block in the months leading up to their disastrous fire.

So, long before Grenfell the Lakanal Inquest established that building regulations were ambivalent enough to allow for flammable materials to be used inappropriately. By devoting resources to the story, BBC London was able to shine a bright light on fire safety, producing in excess of 200 reports across all outlets on the Lakanal fire between 2009 and 2013. Its lead was followed by the *South London Press*, the *Evening Standard* and, on occasions, BBC National News. It was the classic public interest story kept alive by a news provider that had the resources to devote time and energy to keep the story and the debate about fire safety on the public agenda. It's worth recalling what the coroner, Justice Frances Kirkham, recommended to then local government minister, Eric Pickles, in her Rule 43 letter. A ruling that gave Rafael Cervi hope that the deaths of his wife and children might prevent future tragedy.

> It is recommended that your department review Approved Document B … to ensure that it provides clear guidance in relation to Regulation B4 of the Building Regulations, with particular regard to the spread of fire over the external envelope of the building and the circumstances in which attention should be paid to whether proposed work might reduce existing fire protection. (Kirkham, 2013)

The Disruption of the Journalism Ecosystem

The Lakanal House fire came at the very start of the disruptive era provoked by social media. Journalists were increasingly able to connect quickly with local contacts on the ground and feed this information into their newsgathering workflow. The anger in local meetings on the disaster could be fed to journalists quickly. Residents with access to their local representatives could garner information 'the press' found difficult to extract from the press office, and local residents positively encouraged filming in their premises when the local authority banned filming onsite.

The poor quality of FRAs, the questionable *stay put* policy, the cladding, the fire service equipment shortages and shortcomings had all been exposed. It could be reasonably argued that if public interest journalism raised all the obvious safety issues that contributed to the Lakanal fire in 2009 and reported extensively on the findings of the 2013 inquest, this should have identified the need for thorough shifts in public policy. And yet despite all these investigations and the leverage it offered to the Grenfell Action Group to raise important questions, Grenfell Tower still succumbed to disaster in 2017.

In the context of the impact of local journalism on the processes of governance and public responses to disasters, it could be argued that local journalism made serious headway in the aftermath of the Lakanal event. Investigative reporting helped open up the public discourse so the public and public authorities could discuss the plausible remedies to a lax fire safety regime. Reporting made many public officials deeply uncomfortable and demonstrated that local journalism can, to a point, give voice to a disaster-affected community. Nevertheless, no individual has to date been held to account, no one went to jail, although in early 2018 the local authority, the London Borough of Southwark, was heavily fined for breaches of its duty of care as a landlord. The bereaved families also received compensation.

What journalism can do, and did in this case, is ask how and why. What it can't do is force politicians to put things right. The danger is, once the story or spotlight moves on, it becomes difficult to keep using a *culture of exposure* to effect change. All this investigative work was of course done with hindsight at Lakanal, alerted to the dangers of tower block fires post-event. This clearly did offer foresight to the residents involved in the Grenfell Action Group, but it failed to deliver changes at the policy level which could have prevented another disaster. In this sense, even where journalism does work, there are significant limitations to the power of the media to sway the debate onto an arc of justice.

At the local level, many newspapers and their owners are looking back up at the cliff face as one by one they have fallen over the edge. The digital revolution promised renewed possibilities for local activism. Hyperlocal journalism, citizen journalists and bloggers would fill cyberspace with informed discussion and liberating voices. There is some evidence around the Lakanal and Grenfell disasters that this was starting to happen in a rather disjointed way.

In the days and weeks following the Grenfell Tower fire, social media platforms buzzed with comments, questions and demands for direct and meaningful intervention. But how much did all this internet chatter really matter? How did it impact on public discourse? What evidence that a 'Fifth Estate', as Dutton (2009) has called it, was gaining traction to add something new and 'networked' to the public discourse, so often dominated by political elites and the mainstream press? Could social media be an effective replacement for a vigilant local media reporting on such a disaster?

One recent study analysed the type of news discourse that emerged on Twitter in the immediate aftermath of the Grenfell disaster. It argued that despite the vast number of tweets on the subject in the weeks following the disaster, the type of words, topics and subjects raised and those who raised them suggests that the social media outputs often reflect an inclination to support existing institutional views of the subject matter (Barling & Rathnayake, 2018). The most retweeted participant in this research study was Jeremy Corbyn, leader of the UK Labour party. In other words, not yet a replacement, more of a complementary source.

In contrast to the Lakanal fire where voices from the disaster-affected community were amplified by the local media creating a running story and investigating the causes and consequences of the fire, at Grenfell the most powerful impact of the 'Fifth Estate' seems to have been the immediate calls for a public inquiry and placing pressure on Prime Minister Theresa May, who made an exceptionally swift decision to announce an inquiry into the fire at Grenfell. A public inquiry, chaired by Sir Martin Moore-Bick, a former Lord Justice of Appeal, is now fully underway and, at the time of writing, is into its second year.

But there is little systematic evidence that local voices were heard any more clearly following the Grenfell tragedy. This form of platform journalism still lends itself to powerful existing institutional actors. It may be that as more sophisticated approaches evolve to exploit these platforms this will change, but what was clear was that the local reporting lessons of Lakanal had not been learned.

A recent parliamentary inquiry into the Welsh local media landscape confirmed widespread fears about the decline of local journalism, but it also bemoaned the lack of empirical research on the impact of social media on the policymaking process (National Assembly for Wales, 2018). It identified some dysfunction in our understanding of local media and its relationship to democracy during this period of digital transition. A lot of what we are doing is guessing and there is currently no effective means of judging whether new digital sources of information are closing the gap; this has been characterised as the 'News deficit'. In other words, as local papers fail, there is as yet inadequate evidence to show that social media information flows into this gap (Annual Report, 2018). Like all revolutions, we are still in the midst of a process of intense disruption (Rusbridger, 2018). If it remains difficult to comprehensively establish what the sources of digital information are on the ground, it is even more difficult to speculate on how the public discourse might be impacted by these new sources and what this means for decision-makers.

It is nevertheless now possible to see beyond the hyperbole of the falsely declared dawn of social media as an essential tool of accountability in democracy.

This group of disparate participants, a fifth estate, in an evolving ecosystem *could* be a new vehicle for democratic renewal. Networked individuals *can* take advantage of the new methods of distribution to channel robust and critical thinking, even make the voices of ordinary people count. But what matters is what happens in practice.

Conclusions: Quality, Accountability, Public Policy and the Continuing Role of Journalism?

The digital revolution has been very painful. With the loss of so many newspaper titles and paid jobs, it has been hard to see anything but a very bleak future for an alignment of quality and local journalism. It is easy to see how the loss of the local journalists, their newspapers and coverage in the London Borough of Kensington and Chelsea *might* have had an impact on coverage of the concerns so widely circulated by the Grenfell Action Group before the fire that killed 72 people. It often takes a disaster for journalists to recognise that there is actually an important story to be told. It is equally possible that the risks and vulnerabilities that contributed to the disaster at Grenfell may have been missed when there were newspaper titles aplenty. It is easy to be wise after the event.

What on the other hand *is* perhaps unusual about the Grenfell disaster for our understanding of the impact of reporting on public discourse is its proximity in time to another tower block fire disaster, whose causes and consequences *were* covered extensively by the local press. Almost all the issues that contributed to the disaster at Grenfell were well ventilated in

2009 *after* the Lakanal House fire and during the public inquest in 2013 into the deaths at Lakanal House.

The recommendations of that inquest were well reported and indeed dispatched to the secretary of state responsible for the oversight of public housing. The problem is nothing happened. This suggests that it wasn't so much a deficit in journalism, nor the impact of this journalism on public discourse and the extent to which the voices of the adversely affected were put in the public domain, but the quality of policymaking and the decisions taken by government after the facts about what caused the Lakanal fire were well known.

Recent research shows that local voices can provide alternative perspectives in the immediate aftermath of a disaster (Barling & Rathnayake, 2018). These voices are no doubt amplified in the digital ecosystem and across social media. Although it's become abundantly clear that we do need more research to be sure we understand how, or even if, this social media discourse feeds into the policymaking process.

It remains extremely difficult to track the digital newcomers to the news ecosystem and assess the value that they add (Ponsford, 2018). It is even more difficult to assess if they are reducing what we have called here a *News deficit*. More research is clearly needed to map this emerging ecosystem and to properly understand what service it is actually fulfilling in a democratic system which relies on a plurality of views to sustain it and an eclectic mix of outlets to challenge and disseminate those views.

If there are established local media players, there is little doubt that communities at risk from or affected by disaster can have much clearer communication pathways to power. But it is important to recognise that to address hazards and vulnerabilities that cause disaster and support communities in their recovery, needs action by public authorities. The public inquiry into Grenfell is yet to fully report; one of the issues it may well tackle is why matters plainly placed on the public record by journalism were ignored by successive governments. Only this, in truth, can prevent disaster, not journalism alone. This does not mean that local journalism doesn't matter; it does mean, however, in the emerging digital ecosystem we should be careful what we wish for.

References

Annual Report. (2018). *Centre for community journalism*. Cardiff University.
Author Interview with Dominic Ponsford. (2018). Editor-in-Chief Press Gazette UK, December.

Barling, K. (2009, September 28). Who's judging the judge. *BBC online*. Retrieved from https://www.bbc.co.uk/blogs/kurtbarling/2009/09/whos_judging_the_judge.html.

Barling, K. (2017a, June 15, 19:47 BST). *BBC News Channel*. On June 15 I called for a public inquiry on Radio 5 Live and on R4 PM Programme.

Barling, K. (2017b). We've been here before. *British Journalism Review, 28*(3), 37–40.

Barling, K. (2017c). *Investigation of the Lakanal House fire*. Middlesex University. Retrieved from www.lakanalhousefire.co.uk

Barling, K., & Rathnayake, C. (2018). *A topic model analysis approach to understand Twitter public discourse: Grenfell Tower Fire case study*. Research Paper presented at Oxford Internet Institute Conference, September. Retrieved from http://blogs.oii.ox.ac.uk/policy/wp-content/uploads/sites/77/2018/08/IPP2018-Barling.pdf.

Bell, E. (2017, June 25). Grenfell reflects the accountability vacuum left by the crumbling local press. *Guardian*.

Dutton, W. H. (2009). The fifth estate emerging through the network of networks. *Prometheus, 27*(1), 1–15.

Feller, G. (2017). Who demands answers now?. *British Journalism Review, 28*(3), 13–18.

Kirkham. (2013). Retrieved July 1, 2019, from https://www.lambeth.gov.uk/sites/default/files/ec-letter-to-DCLG-pursuant-to-rule43-28March2013.pdf.

Local Media Works. (2019). Retrieved from www.localmediauk.org/research-insight/facts-figures.

Mair, J., Keeble, R., & Fowler, N. (2013). *What do we mean by local? The rise, fall—And possible rise again—Of local journalism*. Bury St Edmunds: Abramis.

Media Reform Coalition. (2019). *Who owns the UK media?*. Goldsmiths University.

Monbiot, G. (2009, November 9). I, too, mourn good local papers, but this lot just aren't worth saving. *Guardian*.

National Assembly for Wales. (2018). *Read all about it: Inquiry into news journalism in Wales*. Culture, Welsh Language & Communication Committee, May.

News Media Association, NMA. (2018). Annual report.

Nielsen, R. K. (Ed.). (2015). *Local journalism: The decline of newspapers and the rise of digital media*. London: I.B. Tauris.

Press Gazette. (2018, October 17). *Google's UK boss says tech giant ready to partner with Government over outcome of Cairncross Review*.

Preston, P. (2017a, September 10). A new silence at the grassroots as papers fade away. *Guardian*.

Preston, P. (2017b, July 2). A functioning local press matters. Grenfell Tower showed us why. *Guardian*.

Rusbridger, A. (2018). *Breaking news: The remaking of journalism and why it matters now*. Edinburgh: Canongate.

CHAPTER 12

Attributes in Community and National News Coverage of the Parkland Mass Shootings

Kyle J. Holody

On February 14, 2018, 17 students and staff members were killed and 17 others were injured when a shooter opened fire in Marjory Stoneman Douglas High in Parkland, Florida. The shooter had formerly attended the high school, had a history of troubling behavior and disciplinary action, and had no apparent motive except to obtain notoriety and express generalized anger, following years of racist, sexual, and aggressive actions toward teachers and other students (Kennedy, 2018; Murphy, 2018; Rozsa, Berman, & Merle, 2018). That the shooter had no specific victims in mind and had perhaps expressed a desire to become "a professional school shooter" (Goldman & Mezzei, 2018) indicate the shootings represent a *rampage* mass shooting, rather than a *mass murder* (shooter is unaffiliated with target location), *targeted* (shooter has specific victims in mind), or other types of mass shootings (McCluskey, 2017; Muschert, 2007; Newman, 2004).

According to a timeline of events published by the *Sun-Sentinel*, near the end of the school day on February 14, the shooter was dropped off at the school by a ride-sharing service, after which at least three staff

K. J. Holody (✉)
Coastal Carolina University, Conway, SC, USA
e-mail: kholody@coastal.edu

J. Matthews, E. Thorsen (eds.), *Media, Journalism and Disaster Communities*, https://doi.org/10.1007/978-3-030-33712-4_12

members either identified the shooter directly or were notified of the presence of a gun on campus, then either took no further action or hid without warning others of the danger (Alanez, McMahon, & Geggis, 2018). A few staff members—Chris Hixon and Aaron Feis—confronted the shooter as soon as they heard of danger, while others—Scott Beigel and Ernie Rospierski—actively sheltered students. All but Rospierski were killed. The shootings lasted only a few minutes; after killing and injuring multiple people in the building, the shooter attempted unsuccessfully to kill people fleeing outside. The shooter then left campus with the escaping crowd, went to nearby businesses and restaurants, and was arrested without incident about an hour later (Alanez, Fleshler, et al., 2018).

The shootings are noteworthy for a variety of reasons. First, the shootings became the deadliest high school shootings in US history. The number of victims and their ages meant the shootings were very likely to attract news attention (Brezenski, 2018; Maguire, Weatherby, & Mathers, 2002; Schildkraut, Elsass, & Meredith, 2018; Silvan & Capellan, 2018). Second, several breakdowns in procedure and communication helped facilitate the shooter's actions. Specifically, the school resource officer had previously been warned the shooter was likely to harm others but concluded there was no threat (Hayes & Perez, 2018). Further, while the shooter had a documented record of behavioral issues and mental health concerns, the shooter was still able to legally purchase guns in Florida (Swisher & McMahon, 2018). Further, miscommunication between responding deputies led to their not entering the building where the shootings were taking place, incorrectly identifying at what parts of campus the shootings were occurring, not knowing if the sounds were gunshots or fireworks, not understanding or sharing with one another that campus security footage they were viewing was on a 20-minute delay, not knowing when the shooter had left campus, and preventing paramedics from entering the building to help survivors even after the shooter had been identified and arrested (Wamsley, 2018).

Finally, the survivors' reactions immediately following the shootings and their activism since also make the Parkland shootings noteworthy. In the days following the shooting, several survivors explained in news interviews that they did not want politicians to respond to the shootings with "thoughts and prayers" but instead with meaningful gun control legislation (Ebbs, 2018; Rossman, 2018). Later, survivors founded or became involved with the #NeverAgain/#EnoughIsEnough advocacy group and the nationwide March for Our Lives demonstration held on March 24, 2018 (Hayes, 2018; Lopez, 2018; Witt, 2018).

Because of the exceptional nature of the Parkland shootings and survivors, it is important to examine how this complex event was presented to news audiences (Park, Holody, & Zhang, 2012). This study examined differences in how the Parkland shootings were covered by community (local and regional newspapers surrounding Parkland) and national newspapers, specifically in terms of the salience and attributes they used to present the shootings. Overall research suggests that local (or *community* in this study) news coverage tends to follow trends set by national organizations (Harry, 2001; Trumbo, 1996). However, research by Holody, Park, and Zhang (2012) and Holody and Daniel (2016) suggest this trend does not always occur for coverage of mass shootings, perhaps because they begin locally before reaching national prominence.

Agenda-Setting: Salience and Attributes

McCombs and Shaw's (1972) agenda-setting theory was utilized to examine if the two newspaper types treated the shootings with different salience, represented by total number of articles about the shootings, the date range over which articles were published, and the average number of words per article. This is especially important for the Parkland shootings, for the Parkland survivors' activism was largely designed to keep the shootings salient, to not let them become just another school shooting (Gurney, Vassolo, Madan, & Smiley, 2018).

In 2004, Chyi and McCombs introduced the news attributes *time* and *space* as common "frames" to be applied to any news issue, thus allowing for the elusive cross-comparison of framing results. Their argument was most studies using framing theory use *frames* (perspectives or storylines for how a news issue should be understood) episodically rather than thematically (Iyengar, 1991). Episodic frames are tied to a particular issue, such as blaming violent video games (e.g., De Venanzi, 2012) or shooters experiencing the "contagion effect" of being inspired by news coverage of mass shootings to seek notoriety through similar actions (DeFoster & Swalve, 2017; Newman, 2004). Thematic frames can be applied to any news issue, such as *conflict* or *David vs. Goliath* conceivably being found in any news story, regardless of topic. Chyi and McCombs (2004) were concerned that the overabundance of episodic frame research meant findings about one news issue could not be compared to another, thus limiting full understanding of how framing and its effects occur. The researchers offered attributes as a form of framing allowing such comparison. This interpretation of the *attribute* concept has since been repeated in several research studies (e.g., Muschert & Carr, 2006; Schildkraut & Muschert, 2013). However, Holody and coauthors (Holody & Daniel, 2016; Park et al., 2012) argue that, while they can be used in conjunction, attributes and frames are inherently distinct concepts. Rather than representing *framing*, these authors argue, attributes are more appropriately included in *second-level agenda-setting*. This study follows the latter line of thinking in its examination of differences in attribute salience between two newspaper types.

Time. Following a range of previous studies (e.g., Cassidy, La France, & Babin, 2018; Holody & Daniel, 2016; Muschert & Carr, 2006; Park et al., 2012; Schildkraut & Muschert, 2013), this study utilized Chyi and McCombs's (2004) *time* attribute of news to examine if news coverage focused on *past* events leading to the shootings, the shootings themselves (*present*), or *future* effects of the shootings. All previous research suggests *present* is the most utilized level in coverage of mass shootings, meaning audiences are most often presented information about shooting events themselves rather than what could lead to or result from shootings.

Space. Chyi and McCombs's (2004) *space* attribute was utilized to examine if news coverage focused on the shooting's effects at *individual*, *community*, *regional*, *societal*, or *international* levels. The present study follows Muschert and Carr's (2006) suggestion to treat any focus on communities outside where the Parkland shootings occurred as *societal*, as opposed to Chyi and McCombs's (2004) *community* suggestion. The majority of previous research suggests news coverage of mass shootings most often utilizes the *societal* level (Cassidy et al., 2018; Chyi & McCombs, 2004; Muschert & Carr, 2006; Park et al., 2012), although two studies found this is not always true (Holody & Daniel, 2016; Schildkraut & Muschert, 2013). Public opinion of shootings was also found by Haider-Markel and Joslyn (2001) to swing between Republicans blaming individual perpetrators and Democrats blaming societal causes, meaning the level used in coverage of mass shootings may relate to public understanding.

Schildkraut and Muschert (2013) predicted that when *space* is used at the *individual* level, news coverage will focus on victims of mass shootings. Hawdown, Oksanen, and Räsänen (2012) found this was true for the 2007 Virginia Tech shootings, though Holody and Daniel (2016) found this was only true for local coverage of the 2012 Aurora shootings. National newspapers focused more on the *shooter*, which is similar to

Dahmen's (2018) findings about the focus of news photographs in shootings coverage.

Attributes combined. A number of studies have examined news coverage by combining the *time* and *space* attributes, similar to Vu, Guo, and McCombs's (2014) concept of *network agenda-setting.* Chyi and McCombs (2004) found national coverage of the 1999 Columbine shootings most used *societal/present* and *community/present.* Muschert and Carr (2006) found national coverage of multiple shootings most used *societal/present* and *community/present.* National coverage most used *societal/present* and *societal/future* for the Virginia Tech shootings (Park et al., 2012); *societal/present* and *individual/present* for the 2012 Sandy Hook shootings (Schildkraut & Muschert, 2013); and *individual/present* and *societal/present* for the 2015 church shootings in Charleston, South Carolina (Cassidy et al., 2018). Finally, national coverage of the Aurora shootings most used *individual/present* and *societal/present*, while local coverage most used *individual/present* and *community/present* (Holody & Daniel, 2016). Overall, the most common attribute combination was *societal/present* until the Aurora shootings in 2012, after which the most common became *individual/present.*

Mention of other shootings. Because the Columbine shootings set the storyline against which all mass shootings since are compared (Schildkraut & Muschert, 2019), the present study also examined if news coverage of the Parkland shootings included references to previous shootings.

Method

Newspaper articles were found by searching for the keywords "Parkland NOT Kentucky" between February 14, 2018 (the date of the shootings), and March 31, 2018 (45 days after the shootings and 7 days after the March for Our Lives protests), in NewsBank for *Broward-Palm Beach New Times* ($N = 4$), *Miami New Times* ($N = 76$), *Sun-Sentinel* ($N = 495$), *Miami Herald* ($N = 707$), and *USA Today* ($N = 91$) and ProQuest for *New York Times* ($N = 215$). A total of 1588 articles were found, from which 25% ($N = 397$) were sampled systematically by choosing every fourth article in reverse date order: *Broward-Palm Beach New Times* ($n = 1$), *Miami New Times* ($n = 19$), *Sun-Sentinel* ($n = 124$), *Miami Herald* ($n = 177$), *USA Today* ($n = 54$), and *New York Times* ($n = 23$). The former four ($n = 321$) represented community coverage and the latter two ($n = 77$) represented national. "Community newspaper" is used here not to describe a community or neighborhood newsletter but, rather, because both local and regional newspapers from within a geographic region were included, "community" refers to newspapers representing the overall community surrounding Parkland, including in Broward, Palm Beach (*Broward-Palm Beach New Times* and *Sun-Sentinel* in Fort Lauderdale), and Miami-Dade (*Miami New Times*, *Miami Herald*) counties, which together make up the Miami metropolitan area.

A final coding category was whether the shootings were central to each news article. This helped ensure only relevant articles were included in the final analysis (Park et al., 2012). As an example, if an article primarily about a sporting event briefly referred to a survivor of the shootings being present, this was coded as "not central." As the shootings were not central for 39 articles, 359 articles (292 community, 67 national) were included in the final results.

Each newspaper article was coded for the following categories. Before coding began, the author and an undergraduate student independently coded a random subsample of 40 articles chosen randomly to establish intercoder reliability. The average Krippendorff's α for nominal-level variables was 0.8463 (minimum: 0.8159), ordinal was 0.9058 (minimum: 0.8814), interval was 1.0000 (minimum: 1.0000), and ratio was 0.8806 (minimum: 0.9078), indicating high reliability for all coding categories.

Salience of shootings. Articles were coded for *original publication* to compare the different levels of newspapers (community, national) and for *number of words* and *publication date* to determine the shootings' salience (i.e., more words indicated greater salience) over time.

Time attribute. Articles were coded for level of the *time* attribute. *Past* indicated focus on events prior to the shootings, such as the shooter's, victims', or survivors' personal histories before February 14, 2018; *present* on what the shooter, victims, survivors, or community members were doing just before, during, or just following the shootings; and *future* on effects of the shootings, including suggestions for solutions, preventions, or actions to be taken.

Space attribute. Articles were coded for level of the *space* attribute. *Individual* indicated focus on the shooter, victims, survivors, or their family members (this could include multiple people, such as a group of survivors); *community* on the school, city, county, or general community where the shootings occurred (e.g., Parkland or Broward County); *regional* on the metropolitan area or state were the shootings occurred (e.g., Miami or Florida); *societal* on other towns, cities, or states in the US or on the country overall; and *international* on countries outside the US or between countries. Articles focusing on certain noteworthy people were coded at other levels of this attribute, even if about specific individuals. For example, an article was coded *community* if it focused on law enforcement or elected officials from Parkland or Broward County, *regional* if on state-elected officials, and *societal* if on President Trump or other federal officials.

If an article was *individual*, it was further coded as focused on the *shooter* or *victims/survivors* (students or school officials who were killed, injured, or present during the shootings, and/or their family members). This was further coded whether the individuals were presented using *sympathy* or *blamed* for the shootings.

Mention of other shootings. Each article was coded for whether it included mentions of other mass shootings in the US. Other types of shootings (such as gang-related or armed robberies) or attacks (such as bombings) were not coded for here, nor were shootings outside of the US.

Results

Salience

Mean number of words. The overall mean number of words per article was 744.86 (SD = 450.294). Community articles had a mean of 730.69 (SD = 459.123) and national 806.61 (SD = 407.032). These were not statistically different: $F(1) = 1.552$, $p = 0.214$.

Date of publication. The overall mean publication date was 17.51 (SD = 12.910) days after the shootings. Community articles had a mean of 17.54 (SD = 12.926) and national 17.39 (SD = 12.937). These were not statistically different: $F(1) = 0.008$, $p = 0.930$.

As Fig. 12.1 demonstrates, community and national newspapers tended to follow the same publication trends, with some evidence community newspapers followed trends from national (i.e., a few spikes in coverage by national newspapers were followed by spikes by community papers). A noticeable difference was that there were no days on which community newspapers published zero articles about the shootings, yet national published zero articles on 10 separate days (including the first two days examined). This could be explained by the fact that the shootings happened in the afternoon, perhaps too late for editions. That no national newspapers were found from the second day is more unusual but may partially be explained the fact that, while there were two articles published on that date each from *USA Today* and *New York Times*, none were sampled.

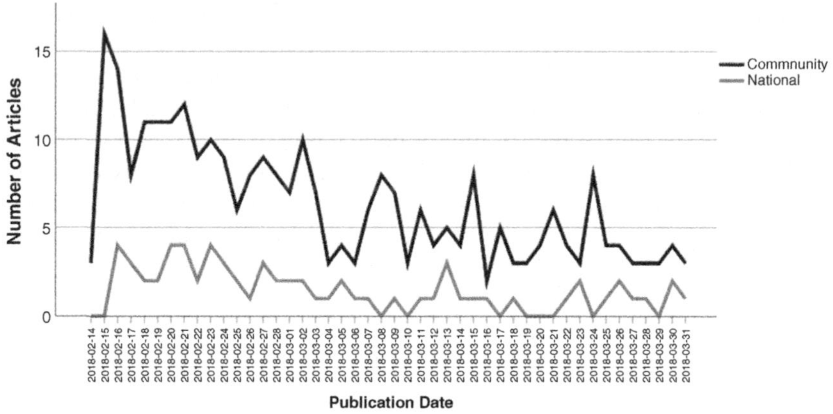

Fig. 12.1 Newspaper publication trends

Further explanation for why only four articles were published in national newspapers on the day following the worst high school shootings in US history can only be speculated upon. Three possible explanations could be that the national newspapers were not able to access local sources as quickly as the community newspapers (Holody et al., 2012) and thus had less information to provide, the national newspapers covered the shootings more in their online coverage than in the print editions analyzed here, or the shootings were not yet considered newsworthy because the shooter had been caught and survivors had not begun their activism.

As Fig. 12.2 demonstrates, there were no significant differences in the number of articles the newspaper types published within different date ranges; each published the highest number within the second date range: $\chi^2(3) = 5.939$, $p = 0.115$.

Attributes

Time. Few newspaper articles utilized the *past* level of the *time* attribute, so *past* and *present* were collapsed into a single category. Community newspapers used *future* in 79.8% of articles; national in 83.6%. These were not statistically different: $\chi^2(1) = 0.498$, $p = 0.480$.

For community articles, 16.9% of *past/present* articles were published in the first date range, 52.5% in the second, 25.4% in the third, and 5.1% in the fourth; 3.9% of *future* were published in the first date range, 40.8% in

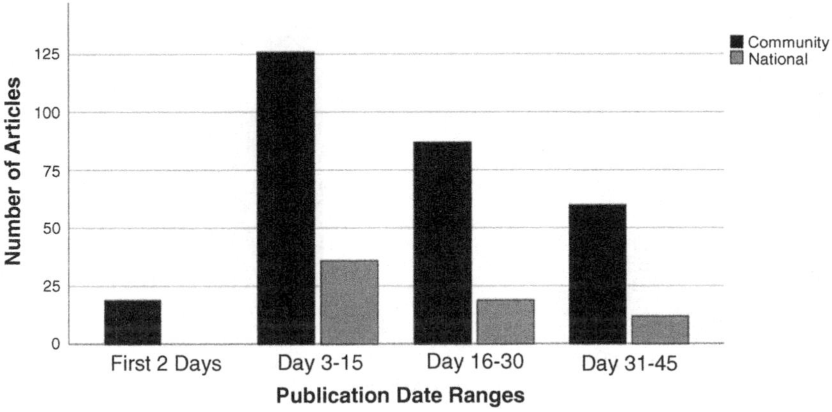

Fig. 12.2 Number of newspaper articles published within different date ranges

the second, 30.9% in the third, and 24.5% in the fourth. These were statistically different: $\chi^2(3) = 22.981$, $p < 0.001$ (see Fig. 12.3).

For national articles, 0% of *past/present* articles were published in the first date range, 72.7% in the second, 9.1% in the third, and 18.2% in the fourth; 0% of *future* were published in the first date range, 50.0% in the second, 32.1% in the third, and 17.9% in the fourth. These were not statistically different: $\chi^2(3) = 2.607$, $p = 0.272$ (see Fig. 12.4).

Space. Few newspaper articles utilized the *international* level of the *space* attribute, so *societal* and *international* were collapsed into a single category. Community newspapers used *individual* in 39.0% of articles, *community* in 23.6%, *regional* in 25.7%, and *societal/international* in 11.6%. National used *individual* in 28.4%, *community* in 3.0%, *regional* in 6.0%, and *societal/international* in 62.7%. These were statistically different: $\chi^2(3) = 90.116$, $p < 0.001$.

For community articles, 7.9% of *individual* articles were published in the first date range, 52.6% in the second, 20.2% in the third, and 19.3% in the fourth; 8.7% of *community* in the first date range, 34.8% in the second, 34.8% in the third, and 21.7% in the fourth; 4.0% of *regional* in the first date range, 41.3% in the second, 32.0% in the third, and 22.7% in the fourth; and 2.9% of *societal/international* in the first date range, 32.4% in the second, 47.1% in the third, and 17.6% in the fourth. These differences approached statistical significance: $\chi^2(9) = 14.782$, $p = 0.097$ (Fig. 12.5).

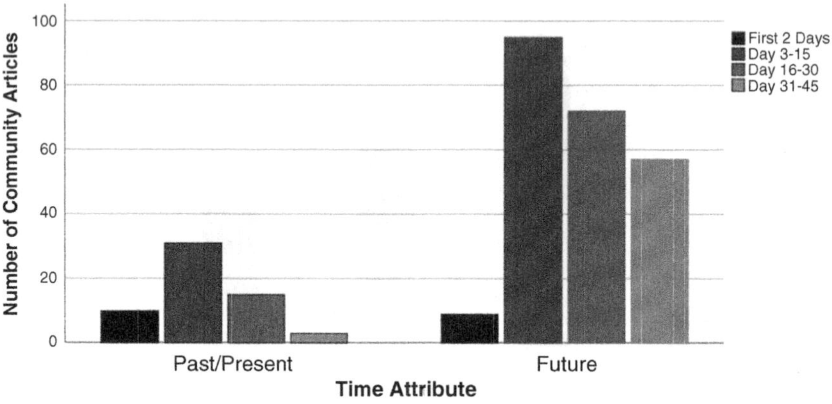

Fig. 12.3 Number of community newspaper articles utilizing past/present or future time attributes

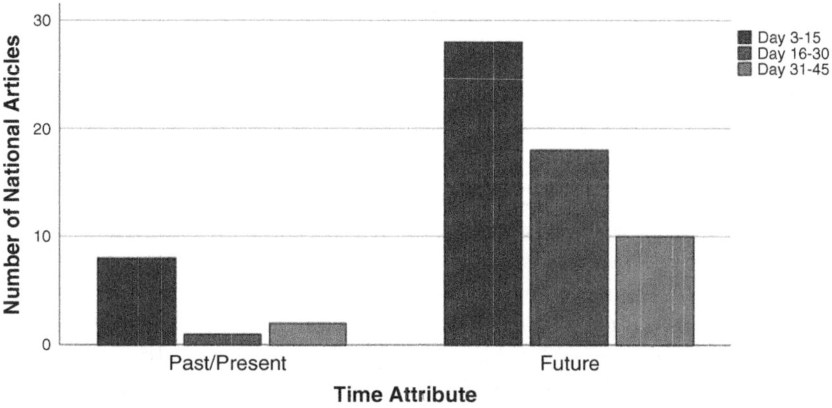

Fig. 12.4 Number of national newspaper articles utilizing past/present or future time attributes

For national articles, 0% of *individual* articles were published in the first date range, 52.6% in the second, 15.8% in the third, and 31.6% in the fourth; 100% of *community* in the second date range; 75.0% of *regional* in the first date range and 25.0% in the second; and 0% of *societal/international* in the first date range, 50.0% in the second, 35.7% in the third, and

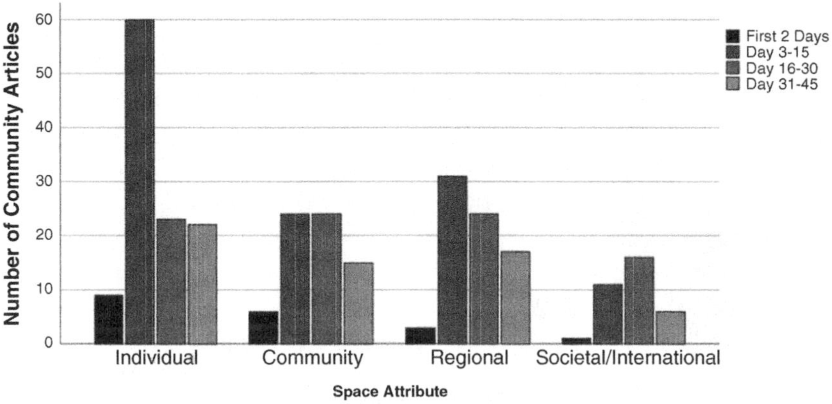

Fig. 12.5 Number of community newspaper articles by space attribute

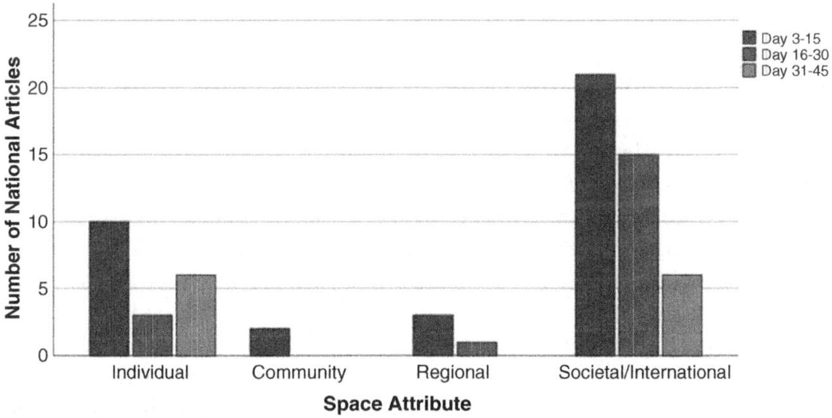

Fig. 12.6 Number of national newspaper articles by space attribute

14.3% in the fourth. These were not statistically different: $\chi^2(6) = 7.054$, $p = 0.316$ (Fig. 12.6).

Of the community articles ($n = 104$) utilizing an *individual* level of the *space* attribute, 26.9% focused on the *shooter* and 73.1% on *victims/survivors*. Of the national ($n = 14$), 7.1% on the *shooter* and 92.9% on the *victims/survivors*. Of the community articles focused on the *shooter* ($n = 28$), 17.9% provided *sympathy*. Of the national ($n = 1$), 100% provided *sympathy*. Of the

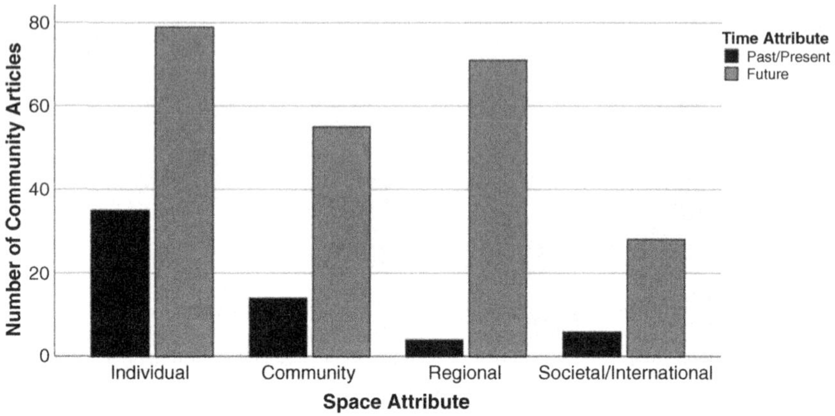

Fig. 12.7 Combinations of space and time attributes for community newspaper articles

community articles focused on the *victims/survivors* ($n = 76$), 98.7% provided *sympathy*. Of the national ($n = 13$), 100% provided *sympathy*. χ^2 tests were not possible because greater than 20% of cells in both comparisons had expected counts less than 5.

Attributes combined. Community articles utilized the time and space attributes together as: *individual/future* (27.1%), *regional/future* (24.3%), *community/future* (18.8%), *individual/past/present* (12.0%), *societal/international/future* (9.6%), *community/past/present* (4.8%), *societal/international/past/present* (2.1%), and *regional/past/present* (1.4%). These were statistically different: $\chi^2(3) = 18.217$, $p < 0.001$ (see Fig. 12.7).

National articles utilized the attributes together as: *societal/international/future* (52.2%), *individual/future* (23.9%), *societal/international/past/present* (10.4%), *regional/future* (6.0%), *individual/past/present* (4.5%), *community/past/present* and *community/future* (1.5% each), and *regional/past/present* (0%). These were not statistically different: $\chi^2(3) = 2.437$, $p < 0.487$ (see Fig. 12.8).

Other Shootings Mentioned

Of the community articles, 20.5% mentioned at least one previous mass shooting. Of the national, 46.3% did. These were statistically different: $\chi^2(1) = 19.052$, $p < 0.001$. Of the community articles mentioning other

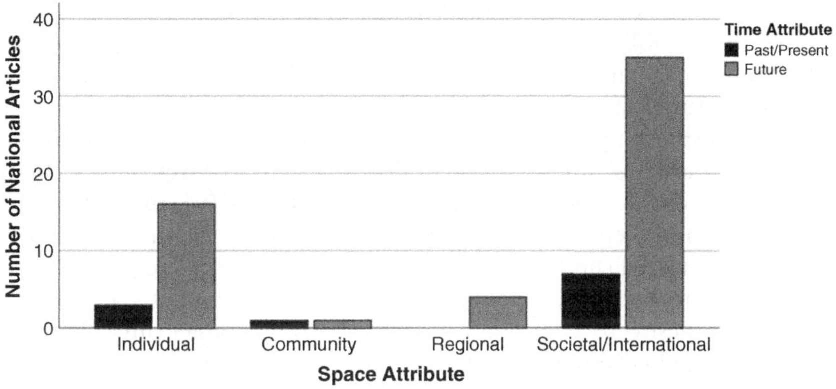

Fig. 12.8 Combinations of space and time attributes for national newspaper articles

shootings, 6.1% were in the first date range, 42.8% in the second, 26.6% the third, and 24.5% the fourth. Of the national, 0% were in the first date range, 52.4% in the second, 28.6% the third, and 19.0% the fourth.

Discussion

Salience

Unlike the coverage of the Aurora shootings, in which national newspapers' mean was higher (Holody & Daniel, 2016), there was no difference between community and national newspapers regarding the mean number of words per article about the Parkland shootings. However, it should be noted that *Miami Times* and *Sun-Sentinel* each published more articles about the shootings than did the national newspapers combined. As Holody et al. (2012) point out, this is likely simply because the shootings were more newsworthy to audiences who live near where the shootings occurred. Community newspapers likely covered the shootings more because of the relevance to the newspapers' audiences. Local and regional newspapers are also less beholden to shifting agendas as compared to national newspapers, which are more likely to cover societal- and international-level news issues because of their intended audiences (Hurka & Neble, 2013).

National news coverage of the Virginia Tech shootings ended just after the 15th day following the shootings (Park et al., 2012; Schildkraut, 2012) and both national and local coverage of the Aurora shootings virtually stopped after the 18th day (Holody & Daniel, 2016). While Columbine coverage lasted longer than this, there was still a noticeable drop in the number of articles published after 15 days (Chyi & McCombs, 2004; Schildkraut & Muschert, 2013). Coverage of the Parkland shootings, however, continued well past the expected 30-day lifetime Chyi and McCombs's (2004) findings suggested mass shootings tend to receive, with an increase in coverage in the second and fourth date ranges. This follows journalists' desire to continue covering shootings longer after the incidents occurred (Dahmen, Abdenour, McIntyre, & Noga-Stryon, 2017). Interestingly, there was no significant difference in how long coverage of the shootings lasted (only the total number of stories differed), despite Holody et al.'s (2012) finding that national coverage of shootings tends to end sooner than local. This is likely because of the newsworthy nature of the survivors, whose activism and engagement with media were especially notable.

Attributes

Both newspaper types used the *future* level of *time* most often, running counter to all previous research finding *present* is the most used level in news coverage of mass shootings (Cassidy et al., 2018; Chyi & McCombs, 2004; Holody et al., 2012; Holody & Daniel, 2016; Muschert & Carr, 2006; Park et al., 2012; Schildkraut & Muschert, 2013). This is perhaps because journalists writing about the Parkland shootings felt that readers already generally understand the all-too-common occurrence of shootings in the US and instead focused on what the effects of this particular shooting will be or how to prevent future shootings. This was also likely bolstered by the continued coverage of the survivors' activism.

While both types most often focused on the *individual* level of *space*, community newspapers did so more often. Further, community newspapers emphasized *community* and *regional* ramifications of the shootings far more often than did national newspapers, which primarily utilized the *societal/international* level. Thus, community newspapers followed the trend found for the Aurora shootings and in some studies of coverage of the Columbine shootings (Holody & Daniel, 2016; Schildkraut & Muschert, 2013), while national newspapers followed the trend for all other mass shootings (Cassidy et al., 2018; Chyi & McCombs, 2004;

Muschert & Carr, 2006; Park et al., 2012). This is likely explained by Park et al.'s (2012) suggestion that local newspapers (or, in this case, local and regional *community* newspapers) are more likely to focus on how the geographic location they are located in is affected, while national newspapers are able to report on shootings from a more removed standpoint (i.e., focusing more on how the shootings affect society overall). For example, multiple articles published in the community newspapers included information about local events organized to garner moral or financial support for the shooting victims and survivors, whereas very few national newspapers did so. The local newspapers also published far more articles about the victims, including the specifics of their lives and memorial services, than did the national newspapers. National newspaper coverage was also more likely to cover people outside or more tangential to the local communities, such as celebrities or politicians who voiced support or donated to relevant causes. In this case, community newspapers focused on the individuals, communities, and regions affected by the shootings because the newspapers' audiences include the specific neighborhoods, cities (Parkland, Fort Lauderdale, and Miami), metropolitan area (Broward, Miami-Dade, and Palm Beach counties), and state (Florida) surrounding where the shootings occurred.

Both newspaper types tended to focus more on the victims and survivors of mass shootings than on the perpetrator, in line with previous trends (Holody & Daniel, 2016; Schildkraut & Muschert, 2013). This is likely for a few reasons. First, the Parkland survivors intentionally and successfully remained in news coverage through their activism. Second, although the shooter's motives remain a mystery and are thus newsworthy, the shooter peacefully gave up to police. Because it is clear there was only one shooter, who was quickly taken into police custody, there was no newsworthy continued threat. Finally, newspapers may have covered the shooter less to deny fame and prevent copycat shootings. It is noteworthy national newspapers covered the shooter far less, but this could be because of their removed relationship with the geographic location where the shootings occurred. In other words, community newspapers likely provided more information because its readership was more likely to include people who knew or had interacted with the shooter.

It is also noteworthy that a small number of community articles expressed *sympathy* for the shooter (17.9%) and *blame* for the victims/survivors (1.3%), while no national articles did so. In almost all cases, these community articles were published in the second date range, when audiences were likely most interested in trying to understand the shooter's history and motives. In other words, the shooter's difficult family life or mental health problems (or supposed bullying at the school) may have provided at least some explanation for why the shootings occurred, in a city recently named one of the safest in the US (Reilly, 2018).

In previous research on news coverage of mass shootings, the most commonly used attribute combination was *societal/present* (Chyi & McCombs, 2004; Muschert & Carr, 2006; Park et al., 2012; Schildkraut & Muschert, 2013), until the Aurora shootings in 2012, after which the most common combination became *individual/present* (Cassidy et al., 2018; Holody & Daniel, 2016). The two shootings for which this change occurred were the Aurora and Charleston, neither of which occurred at a school. In this study, the most commonly used combinations were *individual/future* for community articles and *societal/international/future* for national, meaning both newspaper types most used unique attribute combinations, with more focus on the future.

Other Shootings Mentioned

National articles were significantly more likely to mention previous shootings, while, overall, mentions of previous shootings tended to occur in the second date range. This tendency to mention previous shootings in the days following the shootings can be explained in two ways. First and most simply, most coverage occurred during the second date range. Second, it is likely journalists at both newspaper types mentioned previous shootings early because this was when audiences would likely be most interested in context for what happened. During the very first days of coverage, journalists are likely to focus primarily on the basic facts of what occurred, whereas coverage during the latter date ranges may focus more on how to prevent future shootings. During the second date range, however, journalists may refer to previous shootings to provide more context for readers to understand the shootings beyond their basic facts. This tendency may be less likely for community newspapers—explaining why a lower percentage of community articles mentioned previous shootings—simply because the Parkland shootings remained directly relevant and salient for community audiences longer than for national audiences. Further, national audiences may have less context for what happened specifically, so comparisons to or mentions of previous shootings may have provided further explanation about what happened for audiences less directly affected.

The most commonly mentioned previous shootings were the Sandy Hook shootings, the 2017 Las Vegas shootings, the Virginia Tech shootings, the Columbine shootings, the 2017 Sutherland Springs, Texas church shootings, and (more commonly found in community newspapers) the 2017 Fort Lauderdale-Hollywood International Airport shootings in Broward County, Florida (where the Parkland shootings also occurred) and the 2016 Pulse nightclub shootings in Orlando, Florida. Also noteworthy is that several articles only implied references to these shootings, rather than specifically referring to them (e.g., referring to protecting schools, concerts, movie theaters, or places of worship, without naming specific shootings).

Areas for Future Research

As with any study, the present analysis had limitations and offers suggestions for future research. Coverage of two other recent shootings most often utilized levels besides *societal* (Holody & Daniel, 2016; Schildkraut & Muschert, 2013), so future research should examine if trends have shifted with time or if differences are dependent on shootings' qualities (whether at a school, the ages or types of victims, etc.). Moreover, while the attributes identified by Chyi and McCombs (2004) have frequently been used to analyze coverage of shootings, it is suggested that future research explore other topics (e.g., comparison of news coverage of mass shootings to coverage of forest fires started by arson) to see if the concepts truly can allow comparison across news topics.

Most importantly in regards to this volume about communities affected by disaster and trauma, the news coverage analyzed here suggests that community newspapers were far more likely to cover the individual, community, and regional ramifications of the shootings than were the national newspapers, which focused much more on the societal/international aspects of the shootings, with only some additional focus on individuals, such as the shooter, the victims, and the activism of the survivors. Both newspaper types were also more likely to emphasize the future ramifications of the shootings more than what led to or occurred during the actual event. As suggested by Park et al. (2012), the role of the community newspaper seems to be to explore the specifics of what can be done at the local and state levels of government and at the individual and community levels of society to maintain safety, knowledge, individual rights, and general health of the citizens the newspapers cover and serve. A limitation of

Chyi and McCombs's (2004) *time* and *space* attributes is that they remove rich detail from newspaper coverage, reducing articles to represent only levels of salience instead of treating them as meaningful content (Muschert, 2009; Park et al., 2012). Thus, future research should examine if more detailed characteristics, such as news frames, were distinct between community and national coverage.

Further, the role of the community newspaper seems to be to keep attention on the issue. In this case, because of the survivors' continued activism, the Parkland shootings did not experience the expected timeline or attrition off the news agenda Chyi and McCombs (2004) predicted for mass shootings, at either the community or national level. However, there was still much more coverage in community newspapers, and it is reasonable to predict this trend continued to occur past the dates examined in the present study. While national news coverage may explore the societal explanations for what leads to mass shootings, or their social/international ramifications, the topic is likely replaced as other important issues or events become relevant at the national level. In other words, while "mass shootings" as a concept remain an important, devastatingly tragic issue in the US, the Parkland shootings specifically are much more likely to stay relevant and timely for longer at the community level.

References

Alanez, T., Fleshler, D., Hobbs, S., Huriash, L. J., McMahon, P., O'Matz, M., & Travis, S. (2018, December 28). Unprepared and overwhelmed: Two decades after Columbine and five years after Sandy Hook, educators and police still weren't ready for Parkland. *Sun Sentinel*. Retrieved from https://projects.sun-sentinel.com/2018/sfl-parkland-school-shooting-critical-moments/.

Alanez, T., McMahon, P., & Geggis, A. (2018, June 1). "That's Crazy Boy." School watchman recognized but didn't stop shooter before Parkland massacre. *The Sun-Sentinel*. Retrieved from https://www.sun-sentinel.com/news/crime/fl-florida-school-shooting-campus-monitor-20180619-htmlstory.html.

Brezenski, T. F. (2018). Inside the 23rd Congressional District (FL) Gun Violence Task Force: Real-time crisis policymaking in the wake of the Marjory Stoneman Douglas school shootings. *Journal of Multidisciplinary Research, 10*(1/2), 35–49.

Cassidy, W. P., La France, B. H., & Babin, S. (2018). Routine adjustments: How journalists framed the Charleston shootings. *International Journal of Communication, 12*, 4669–4688.

Chyi, H. I., & McCombs, M. E. (2004). Media salience and the process of framing: Coverage of the Columbine school shootings. *Journalism & Mass Communication Quarterly, 81*(1), 22–35. https://doi.org/10.1177/107769900408100103

Dahmen, N. S. (2018). Visually reporting mass shootings: US newspaper photographic coverage of three mass school shootings. *American Behavioral Scientist, 62*(2), 163–180. https://doi.org/10.1177/0002764218756921

Dahmen, N. S., Abdenour, J., McIntyre, K., & Noga-Stryon, K. E. (2017). Covering mass shootings: Journalists' perceptions of coverage and factors influencing attitudes. *Journalism Practice, 12*(4), 456–476. https://doi.org/10.1080/17512786.2017.1326832

DeFoster, R., & Swalve, N. (2017). Guns, culture or mental health? Framing mass shootings as a public health crisis. *Health Communication, 33*(10), 1–12. https://doi.org/10.1080/10410236.2017.1350907

De Venanzi, A. S. (2012). School shootings in the USA: Popular culture as risk, teen marginality, and violence against peers. *Crime, Media, Culture, 8*(3), 261–278. https://doi.org/10.1177/1741659012443233

Ebbs, S. (2018, February 15). Survivors of Florida high school shooting call for action on gun control. *The Washington Post*. Retrieved from https://abcnews.go.com/Politics/survivors-florida-high-school-shooting-call-action-gun/story?id=53111278.

Goldman, A., & Mezzei, P. (2018, February 15). YouTube comment seen as early warning in shooting left little for F.B.I. to investigate. *The New York Times*. Retrieved from https://www.nytimes.com/2018/02/15/us/politics/nikolas-cruz-youtube-comment-fbi.html.

Gurney, K., Vassolo, M., Madan, M. O., & Smiley, D. (2018, February 18). How teachers and students became first responders—And heroes—In school shooting. *The Miami Herald*. Retrieved from https://www.miamiherald.com/news/local/community/broward/article200597734.html.

Haider-Markel, D. P., & Joslyn, M. R. (2001). Gun policy, opinion, tragedy, and blame attribution: The conditional influence of issue frames. *Journal of Politics, 63*(2), 520–543. https://doi.org/10.1111/0022-3816.00077

Harry, J. (2001). Covering conflict: A structural-pluralist analysis of how a small-town and a big-city newspaper reported an environmental controversy. *Journalism & Mass Communication Quarterly, 78*(3), 419–436. https://doi.org/10.1177/107769900107800302

Hawdown, J., Oksanen, A., & Räsänen, P. (2012). Media coverage and solidarity after tragedies: The reporting of school shootings in two nations. *Comparative Sociology, 11*(6), 1–30. https://doi.org/10.1163/15691330-12341248

Hayes, C. (2018, February 17). Emma Gonzalez survived the Florida shooting. Now she's taking on Trump and the NRA. *USA Today*. Retrieved from https://www.usatoday.com/story/news/2018/02/17/student-emma-gonzalez-school-shooting-gives-passionate-speech-against-g/348357002/.

Hayes, C., & Perez, M. (2018, February 22). 'It's devastating': Cop stayed outside during Florida massacre while students died. *USA Today*. Retrieved from https://www.usatoday.com/story/news/2018/02/22/officer-florida-high-school-shooting-sat-outside-gunfire-sounding-he-should-have-stopped/365122002/.

Holody, K. J., & Daniel, E. S. (2016). Attributes and frames of the Aurora shootings: National and local news coverage differences. *Journalism Practice, 11*(1), 80–100. https://doi.org/10.1080/17512786.2015.1121786

Holody, K. J., Park, S.-Y., & Zhang, X. (2012). Racialization of the Virginia Tech shootings: A comparison of local and national newspapers. *Journalism Studies, 14*(4), 568–583. https://doi.org/10.1080/1461670X.2012.726499

Hurka, S., & Neble, K. (2013). Framing and policy change after shooting rampages: A comparative analysis of discourse networks. *Journal of European Public Policy, 20*(3), 390–406. https://doi.org/10.1080/13501763.2013.761508

Iyengar, S. (1991). *Is anyone responsible? How television frames political issues*. Chicago, IL: University of Chicago Press. https://doi.org/10.7208/chicago/9780226388533.001.0001

Kennedy, K. (2018, February 15). Here's what we know about Nikolas Cruz, the Florida school shooting suspect. *The Boston Globe*. Retrieved from https://www.boston.com/news/national-news/2018/02/14/nikolas-cruz-parkland-florida-school-shooting-suspect.

Lopez, G. (2018, March 26). It's official: March for our lives was one of the biggest youth protests since the Vietnam War. *Vox*. Retrieved from https://www.vox.com/policy-and-politics/2018/3/26/17160646/march-for-our-lives-crowd-size-count.

Maguire, B., Weatherby, G. A., & Mathers, R. A. (2002). Network news coverage of school shootings. *The Social Science Journal, 39*(3), 465–470. https://doi.org/10.1016/S0362-3319(02)00201-X

McCluskey, M. (2017). *News framing of school shootings: Journalism and American social problems*. Lanham, MD: Lexington.

McCombs, M. E., & Shaw, D. L. (1972). The agenda-setting function of the mass media. *Public Opinion Quarterly, 36*, 176–187. https://doi.org/10.1086/267990

Murphy, P. P. (2018, February 18). Exclusive: Group chat messages show school shooter obsessed with race, violence and guns. *CNN*. Retrieved from https://edition.cnn.com/2018/02/16/us/exclusive-school-shooter-instagram-group/index.html.

Muschert, G. W. (2007). Research in school shootings. *Sociology Compass, 1*(1), 60–80. https://doi.org/10.1111/j.1751-9020.2007.00008.x

Muschert, G. W. (2009). Frame-changing in the media coverage of a school shooting: The rise of Columbine as a national concern. *Social Science Journal, 46*(1), 164–170. https://doi.org/10.1016/j.soscij.2008.12.014

Muschert, G. W., & Carr, D. (2006). Media salience and frame changing across events: Coverage of nine school shootings, 1997–2001. *Journalism & Mass Communication Quarterly, 83*(4), 747–766. https://doi.org/10.1177/107769900608300402

Newman, K. (2004). *Rampage: The social roots of school shootings*. New York: Basic Books.

Park, S.-Y., Holody, K. J., & Zhang, X. (2012). Race in media coverage of school shootings: A parallel application of framing theory and attribute agenda-setting. *Journalism & Mass Communication Quarterly, 89*(3), 475–494. https://doi.org/10.1177/1077699012448873

Reilly, K. (2018, February 15). The location of the Florida high school shooting was recently named the safest city in the state. *Time*. Retrieved from http://time.com/5159277/parkland-florida-high-school-shooting-safest-city/.

Rossman, S. (2018, May 18). Parkland students said, 'We will be the last mass shooting.' Now, they grieve with Texas. *USA Today*. Retrieved from https://www.usatoday.com/story/news/nation-now/2018/05/18/texas-school-shooting-parkland-students-now-grieve-santa-fe-texas/623265002/.

Rozsa, L., Berman, M., & Merle, R. (2018, February 18). Accused South Florida school shooter confessed to rampage that killed 17 people, police say. *The Washington Post*. Retrieved from https://www.washingtonpost.com/news/post-nation/wp/2018/02/15/florida-school-shooting-suspect-booked-on-17-counts-of-murder-premeditated/.

Schildkraut, J. (2012). Media and massacre: A comparative analysis of the reporting of the 2007 Virginia Tech shootings. *Fast Capitalism*. Retrieved from http://www.uta.edu/huma/agger/fastcapitalism/9_1/schildkraut9_1.html.

Schildkraut, J., Elsass, H. J., & Meredith, K. (2018). Mass shootings and the media: Why all events are not created equal. *Journal of Crime and Justice, 41*(3), 223–243. https://doi.org/10.1080/0735648X.2017.1284689

Schildkraut, J., & Muschert, G. W. (2013). Violent media, guns, and mental illness: The three ring circus of causal factors for school massacres, as related in media discourse. *Fast Capitalism*. Retrieved from http://www.uta.edu/huma/agger/fastcapitalism/10_1/schildkraut10_1.html.

Schildkraut, J., & Muschert, G. W. (2019). *Columbine, 20 years later and beyond: Lessons from tragedy*. Santa Barbara, CA: Praeger.

Silvan, J. R., & Capellan, J. A. (2018). The media's coverage of mass public shootings in America: Fifty years of newsworthiness. *International Journal of Comparative and Applied Criminal Justice, 43*(1), 77–97. https://doi.org/10.1080/01924036.2018.1437458

Swisher, S., & McMahon, P. (2018, February 15). Nikolas Cruz passed background check, including mental health question, to get AR-15 rifle. *The Sun-Sentinel*. Retrieved from https://www.sun-sentinel.com/local/broward/parkland/florida-school-shooting/fl-florida-school-shooting-guns-20180215-story.html.

Trumbo, C. (1996). Constructing climate change: Claims and frames in US news coverage of an environmental issue. *Public Understanding of Science, 5*(3), 269–283. https://doi.org/10.1088/0963-6625/5/3/006

Vu, H. T., Guo, L., & McCombs, M. E. (2014). Exploring "the world outside and the pictures in our heads": A network agenda-setting study. *Journalism & Mass Communication Quarterly, 91*(4), 669–686. https://doi.org/10.1177/1077699014550090

Wamsley, L. (2018, February 26). Broward sheriff under scrutiny for handling of Parkland shooting. *NPR: The Two-Way*. Retrieved from https://www.npr.org/sections/thetwo-way/2018/02/26/588900401/broward-sheriff-under-scrutiny-for-handling-of-parkland-shooting.

Witt, E. (2018, February 19). How the survivors of Parkland began the never again movement. *The New Yorker*. Retrieved from https://www.newyorker.com/news/news-desk/how-the-survivors-of-parkland-began-the-never-again-movement.

CHAPTER 13

Informing Refugee Communities in Greece: What Is Possible Within the Parameters of the Humanitarian Structure?

Victoria Jack

When communities are affected by an earthquake or landslide, they may look to familiar and trusted local media outlets for information. But what happens when you're a refugee displaced in a foreign country? The local media don't speak your language and your community isn't part of their imagined audience. The potential role of local media in addressing the information and communication needs of refugees is further complicated by the highly political and dynamic nature of contemporary displacement crises. The information refugees seek to access often is not only procedurally complex but politically sensitive. Meeting these needs is particularly challenging at a time when policy and legal responses to displacement are characterised by securitisation in addition to—or even rather than—the protection of rights. Owing to these distinct challenges, it is important to explore the role of local media in displacement crises separately to other types of disasters.

V. Jack (✉)
University of New South Wales, Sydney, NSW, Australia

J. Matthews, E. Thorsen (eds.), *Media, Journalism and Disaster Communities*, https://doi.org/10.1007/978-3-030-33712-4_13

In the past few decades, there has been a growing discourse about the need for humanitarian response to not only address physical needs—such as food, water and shelter—but to also provide information and engage in two-way communication with crisis and disaster-affected communities. Communication with Communities is an emerging field of humanitarian response, charged with ensuring that people affected by crisis can access the information they need and communicate with the actors providing humanitarian assistance. There is a lack of research about Communication with Communities, including the role that media initiatives could play in providing vital information and giving voice to communities. This chapter begins to address this gap by considering if and how local media initiatives can operate within the parameters of the humanitarian structure. To do so, the chapter draws on ethnographic fieldwork conducted during the so-called European Union (EU) Refugee Crisis.

In 2015, more than a million refugees and migrants arrived at Europe's borders. They came primarily from Syria, Afghanistan and Iraq. The vast majority arrived by boat from Turkey to one of four nearby Greek islands—Lesvos, Kos, Chios or Samos—before attempting to travel onwards. At the height of the crisis, thousands of people arriving each day were in need of information that could help them access humanitarian assistance and navigate EU asylum procedures. Drawing on the researcher's experience of working for a humanitarian information project called *News That Moves* in Greece during the crisis, this case study facilitates reflection as to the role that such initiatives can play in protecting the right to information in situations of displacement.

Access to Media and Information in Refugee Camps

The conceptual framework of humanitarian response to situations of forced migration is founded on the principle of protection. The United Nations Refugee Agency (UNHCR) and its implementing partners seek to provide international protection to refugees, which is envisioned as involving both physical security and legal security (UNHCR, 2011). UNHCR takes a holistic approach to international protection, which involves ensuring that rights contained within international refugee instruments are respected and that the fundamental rights embodied in other international conventions are also observed (UNHCR, 2006b). Therefore, "the phrase 'international protection' covers the gamut of activities through which refugees' rights are secured" (UNHCR, 1999, p. 3). Of particular relevance to this discussion are the rights outlined in Article 19 of the Universal Declaration of Human Rights:

> Everyone has the right to freedom of opinion and expression; this right includes the right to hold opinions without interference and to seek, receive and impart information and ideas through any media and regardless of frontiers. (UNHCR, 2006a, p. 58)

UNHCR's global reference guide for the protection of refugees in camps and settlements states that refugees should have access to "accurate, reliable, regular and up-to-date information" about matters such as the assistance and programmes available in the camp, the situation in their communities of origin, their rights and obligations, and explanations for the decisions made by humanitarian organisations (UNHCR, 2006a, p. 58). Yet, while the logic of protection enables the humanitarian government to provide assistance, it simultaneously enables the management and policing of refugee populations. As Agier (2011, p. 4) argues, "There is no care without control." The power assigned to actors involved in humanitarian response enables these actors to make decisions about what information is provided to refugees—whose futures hang in the balance—and what is withheld.

Although the literature points to the importance of information access for refugees, there had been minimal scholarly exploration of the ways that refugees access and use media prior to the EU Refugee Crisis. In 2015–2016, the visibility of smartphone use by refugees travelling to and in Europe fuelled a spike of interest in digital connectivity. In a study based on interviews with Syrian and Iraqi refugees in France, the main finding of Gillespie, Osseiran, and Cheeseman (2018, p. 15) was that Wi-Fi and battery-charging resources are "the very first things refugees look for" upon arrival in Europe. Digital connectivity is linked to a range of affordances, including keeping in touch with family and friends and accessing information and resources that facilitate onward travel (Alencar, Kondova, & Ribbens, 2019; Gillespie et al., 2018; Kutscher & Kreß, 2018; Latonero & Kift, 2018). Despite the potential utility of smartphones and social media, the extent to which refugees use and rely on smartphones may be constrained both by limited opportunities for connectivity and by concerns about government surveillance (Alencar et al., 2019). Such concerns lead Syrian refugees to prefer social media information obtained from existing social ties rather than official sources (Dekker, Engbersen, Klaver,

& Vonk, 2018). It is noteworthy that existing studies primarily involve the participation of Syrian refugees. It is unclear whether these patterns would apply to refugees from other countries of origin, where lower levels of education and digital literacy may apply.

Despite growing academic interest in smartphone use following the EU Refugee Crisis, there remains a lack of research concerning the role that news media might play in addressing information needs in a camp environment. The small number of scholarly studies that explore access to news media in situations of displacement describes a lack of access to regular and reliable sources (Gillespie et al., 2016, 2018; Jack, 2017; Kivikuru, 2013; Robinson, 2004). The limited range of news media available to refugees in camps is generally characterised by a lack of direct relevance to the refugee experience. There are some instances, including camps in Palestine and Thailand, where community media platforms facilitate the provision of information about the situation inside the camps. However, it is typical even in camp environments where there are community media that such media do not report in a manner that is directly relevant and accessible to the refugees themselves. Instead:

> The messages—often highly political—mainly focus abroad, trying to convince the rest of the world about the need for peace and democracy. These media rarely transmit stories about the day-to-day life in the camps and their audiences are outside the camps. Thus, although media content is produced on the camps, the focus is actually on the world outside. (Kivikuru, 2013, p. 37)

Existing studies do not focus in detail on examining the potential of news initiatives designed specifically for refugees and migrants or the extent to which such content can address the needs of the intended audience.

Despite the lack of research on the role of news media in refugee camps, the literature reveals some synthesis between the role of Communication with Communities—as outlined earlier in this section—and the role of local and community media. Local media are charged with providing information to local communities and amplifying voices from within those communities. The rationale for this position is that information can enable the community to make informed decisions and that enabling them to express their opinions can help hold powerful individuals and institutions to account (Franklin, 2006). The media thus play a role in "the local democratic process as investigators for, and informers of, the public"

(Williams, 2006, p. 83). These objectives seem to correspond with the humanitarian duty to ensure respect for the right to freedom of opinion and to seek information "through any media and regardless of frontiers" (UNHCR, 2006a, p. 58). But how do these ideals play out in practice? This chapter reflects on the activities of a humanitarian information project that operated during the EU Refugee Crisis, in order to consider the role that such initiatives can play in protecting the right to information in situations of humanitarian crisis.

The EU Refugee Crisis

"We do not care where we sleep and what we eat. We just want to know what is going to happen to us and when." These words were uttered by a 40-year-old Kurdish Iraqi male in April 2016—the month after the EU-Turkey deal came into effect. The man was staying at an unofficial encampment known as "Stonehouse", at the port of Piraeus near Athens. Thousands of others were also living in poor conditions, as Greece struggled to establish the facilities needed to accommodate those who were now stuck on the country's mainland. These circumstances were dramatically different to those of the previous year, when refugees and migrants—mainly from Syria, Iraq and Afghanistan—had been able to travel onwards from Greece to destinations such as Germany, France and Sweden. The EU's Dublin system, which obliges refugees to seek asylum in the first port of irregular entry to the region, put enormous pressure on Greece to accommodate and process the huge numbers of arrivals. By Spring 2015, Greece was unable to cope and started waving people on, creating "secondary movements" through the Balkan route to Western Europe. However, by late 2015, countries were re-establishing restrictions as the EU moved to regain control over its internal borders.

Following the introduction of the EU-Turkey deal in March 2016, the Greek authorities, in collaboration with other EU actors, scrambled to draft and implement confusing new policies around asylum, relocation and deportation. Refugees were now confined to camps in Greece and—their savings having run out—they became increasingly dependent on humanitarian aid. Previously, refugees relied heavily on social media sites dedicated to seeking asylum in Europe, where other refugees would post useful information about travelling onwards from Greece and where to access services. Now the questions to which refugees needed answers shifted from relatively simple concerns about directions and access to basic aid, to complex legal matters and procedures to which the answers did not yet exist or were unclear even to asylum law experts. In this new environment, refugees needed access to reliable information that could inform their decisions about legal procedures and access to assistance—and a way to project their voices to the humanitarian organisations and Greek authorities that made decisions about their lives.

News That Moves

News That Moves sought to meet the information needs of refugees affected by the EU Refugee Crisis, as well as to elevate their voices into the public domain. The media initiative was launched in late 2015 by Internews, a non-government organisation that is a pioneer in the field of humanitarian communications. A team of Internews staff produced clear, factual and useful information—"news-you-can-use"—designed to meet the needs of refugees arriving in Greece and travelling along the Balkans route to Western Europe. While the refugees did not share a specific locale—they represented a disaster community which shared a common experience and an accompanying set of specific information and communication needs. The main platform for distribution was the News That Moves website—and later Facebook—because it was important to provide a service that could be accessible to refugees as they travelled. A partnership with Translators Without Borders enabled the provision of information in four languages—English, Greek, Arabic and Farsi/Dari. I worked as the project's Humanitarian Data Manager throughout 2016. The reflections offered here are drawn from my ethnographic field notes.

Displaced communities are often unable to access trusted sources of local media from their place of origin. Meanwhile, media in the host community do not meet the needs of new arrivals because these "outsiders" are not part of the media's imagined audience. For these reasons, it was necessary to establish a media and information service designed to meet the particular needs of refugees displaced in Greece and other parts of the EU. However, the establishment of a new initiative presents its own challenges—not least getting word out to affected communities and building their trust in the platform. A study of the types of media used by Haitians following the 2010 earthquake suggests that crisis-affected communities are more likely to turn to trusted pre-existing channels than to take up newly available sources that are not part of their pre-crisis information repertoire (Sommerfeldt, 2015). Using an internet browser to access news was not common practice among refugees in Greece, but many used smartphones to access updates on social media. For this reason, a decision was made to cross-post all News That Moves website content to dedicated Arabic and Farsi language Facebook pages. To bring News That Moves to the attention of the target audiences, Internews promoted the service in posts on other social media sites used by refugees travelling to and in Europe. Additionally, Internews ran regular promotions—or "boosts"—to target specific demographics of Facebook users (e.g. Arabic or Farsi speakers currently located in Greece). These tactics helped to grow the total audience of the Arabic and Farsi News That Moves Facebook pages to more than 50,000 (34,000+ Arabic and 18,000+ Farsi). The English and Greek websites were keenly read by humanitarians working in Greece, who were eager to access reliable news about the unfolding circumstances of the crisis.

The project sought to establish trust by building a reputation as a credible source of relevant information. A team of Refugee Liaison Officers gathered feedback from refugees in official and unofficial encampments on a daily basis, which enabled the team to develop a sound understanding of refugees' concerns as they changed, and then to address these concerns in News That Moves content. News That Moves staff documented and analysed Facebook engagement—in the form of comments, questions, shares and likes—to obtain further insights into the information needs, preferences and concerns of the target audiences. Total audience engagement between April and September 2016 included 6642 comments (including questions), 1791 shares and 257,938 likes.

Dedicated engagement campaigns were undertaken to generate timely feedback on breaking stories affecting refugees in one or more of the 40+ encampments located across Greece. These campaigns helped the News That Moves team cover events happening in sites where we could not physically be present. The campaigns also enabled the team to solicit feedback from sites across Greece, which enhanced our capacity to identify topics of relevance to refugees in particular sites and across sites. One example of a site where the team could not physically be present at the time of an important development was during the shutdown of an encampment at Idomeni, where more than 10,000 people had gathered near the Macedonia border in the hope of being allowed to continue their journey. In May 2016, word spread that the Greek government was set to conduct an evacuation—but official information was scant, and journalists were prevented from accessing the site during the operation. To obtain information, News That Moves reached out directly to refugees at Idomeni through a Facebook post:

> Have you been relocated from Idomeni to another camp? Tell us what the conditions are like where you are staying now. How do you feel about the move? What are the accommodation and services like? Do you have access to food, water, wifi, etc.? Let us know which camp you're in and what is your situation.

The post attracted 251 comments (as well as 724 likes), which were analysed by the team and used to inform our online coverage. The feedback enabled News That Moves to provide information about how and where refugees and migrants were being moved and to give voice to their concerns about the substandard conditions of the new camps.

Over the months, News That Moves covered a great range of topics that reflected the changing context of the crisis and, consequently, the changing needs of refugees and migrants. Before the borders closed, the information needs of the moving populations focused on access to services along migration routes. "Is there somewhere we can sleep?" "Am I able to get some food?" "How do I get a *kharti* [registration paper]?" "Where can I buy a ticket for the ferry from Lesvos to Athens?" News That Moves was usually able to provide actionable answers to these questions. The service enhanced the audience's capacity for agency by enabling them to make decisions and to access food, shelter and other services as they travelled onwards towards their destination.

Following the border closures and introduction of the EU-Turkey deal, it became more difficult and politically sensitive to disseminate the types of information refugees and migrants were seeking. It took months for the Greek authorities to set up procedures that enabled refugees and migrants now stuck in Greece to register for asylum, relocation and family reunification. News That Moves worked to inform refugees and migrants of their options for the future, as well as their rights and restrictions in Greece, and the availability of services and assistance. However, frustrated by the delays and worried that they may wait a year or more only to be told that their application would be denied, many considered attempting to travel onwards despite the border restrictions—often with the help of a smuggler. In the increasingly securitised environment following the EU-Turkey deal, there emerged a fission between what refugees and migrants wanted to know and what the humanitarian government would permit them to know.

Discussion

The conceptual framework of humanitarian response to situations of forced migration is, as outlined earlier in this chapter, founded on the principle of protection. This includes the protection of rights—such as freedom of expression and to receive information through any media and regardless of frontiers. In theory, protection and access to media are united by a mutual emphasis on enabling communities to make informed decisions and amplifying their voices. But on closer inspection, it is when an attempt is made to apply these ideals that the complexities of the relationship between humanitarian protection and access to information become apparent.

The circumstances of displacement produce a range of protection concerns that humanitarian agencies seek to manage and mitigate as part of the humanitarian response. Refugees are, by virtue of their being refugees, in an extremely vulnerable position. In Greece, refugee communities were exposed to people smugglers, antisocial behaviour and physical violence, protests, attacks from far-right groups and outbreaks of disease, among other threats. People are frustrated and tensions run high. In such an environment, providing certain types of information can present additional risks—while failing to provide other types of information can exacerbate already-present risks—more than is the case for the average reader of a community or local news website in a non-crisis environment. An example of the tragic consequences that the dissemination of information can cause is the drowning of three Afghans, who tried to cross the icy cold waters of the River Suva on the Greek border in March 2016. They did so shortly after a group of activists distributed a leaflet in Idomeni camp, encouraging refugees to join a mass border crossing and falsely claiming that "the river is dry".

The risks and vulnerability associated with displacement mean that the traditional practices of community and local media cannot be simply transferred from non-crisis settings in which they are normally deployed to the vastly different environment of a humanitarian crisis. In cases where media initiatives are active in humanitarian crises, their journalists and contributors are faced with deciding whether to provide or withhold information—knowing that in the high stakes environment in which they are working information could save a life or put a life in danger. The decisions of journalists and contributors to initiatives are not solely their own, however, as the host government and humanitarian agencies seek to influence decisions about what information is provided. The provision of information to communities could, if ill-considered, compromise humanitarian efforts to protect rights. Yet, a lack of access to information can be a breach of rights and thus an important protection issue in itself. Refugees in Greece were exposed to "a great deal of disinformation (lies), misinformation (inaccurate information), false rumours, and conspiracy theories via social media networks… making their journeys even more precarious" (Gillespie et al., 2018, p. 3). Owing to the serious consequences of restricting access to information, the logics invoked by governmental and humanitarian actors when assuming the right to withhold information warrant examination.

An occasion when information was withheld from the refugee communities in Greece, citing a protection logic about the possible harms that could be caused, was during the Greek government's exercise to pre-register asylum applications in 40-plus official sites on the Greek mainland. Refugees were told that teams from the Greek Asylum Service would visit each site sometime between the end of May and the end of July. Refugees were not told the date that each site would be visited. Governmental and humanitarian actors expressed a concern that if the precise schedule was made available, refugees who discovered that their site was not to be pre-registered until late July would flock to sites set to be pre-registered in late May or early June in order to speed up their registration process. There was concern that the gathering of large crowds of people desperate to pre-register for asylum after months of delays could create physical danger and compromise the successful execution of the exercise. Owing to these requests, News That Moves decided to not publish the pre-registration schedule. Refugees and migrants were eager to know when their site would be pre-registered because they wanted to ensure that they would be present on the day the pre-registration exercise was conducted at their site. There were many reasons why refugees and migrants travelled away from their site for several hours or days at a time—such as to pursue work opportunities, to attend medical appointments and to visit friends staying in other sites. Indeed, several refugees told News That Moves that they were left in limbo after missing out on their opportunity to pre-register because they were unintentionally absent on the day the Greek Asylum Service turned up.

The rationale for withholding information about the pre-registration schedule from refugees suggests that governments and humanitarian agencies may view the provision of information to refugees as a threat to the extent to which they can be managed. Correspondingly, the most recent version of the UNHCR Resettlement Handbook suggests that while refugees have a right to information "in principle", when deciding what to communicate "UNHCR needs to weigh its own interests... against the refugee's legitimate interests" (2011, p. 125). UNHCR's interests have included the promotion of repatriation and local integration policies, and efforts to prevent refugees from undermining resettlement processes through tactics such as fraudulent resettlement claims (Saltsman, 2010) and "asylum shopping" (Saltsman, 2011). Previous studies have highlighted, for example, the tendency of UNHCR to withhold information about resettlement processes and eligibility criteria as a way of insulating these processes from so-called refugee cheating (Jansen, 2008). It is evident, then, that there are instances in which the provision of information is viewed as a potential threat to the extent to which refugee populations can be managed.

Paradoxically, my research suggests that a failure to communicate also presents a threat to the humanitarian capacity to assist and control. For example, if refugees are not adequately informed about the legal options available to them—and how to access these options—then they may feel that there is no alternative but to travel onwards with the help of a smuggler. Refugees lost faith in the possibility that the asylum process was a reliable and timely solution to their displacement, one Syrian man told News That Moves, after they did not receive a text message from the Greek Asylum Service regarding the date of their next appointments in the asylum process. "The texts for the next appointment after pre-registration haven't arrived yet. Many people went with smugglers," he explained. This story demonstrates that cases of "refugee cheating" (e.g. travelling onwards with a smuggler) should not be interpreted as the cause of failures within a humanitarian system, but as a response to already-present failures—such as failing to meet the information needs of refugees in a manner that gives them faith in the capacity of humanitarian processes (e.g. asylum) to present a solution to their predicament. Such failures are constituted at least in part by communication gaps that diminish refugees' faith in humanitarian procedures and give rise to coping tactics that involve working around rather than within humanitarian structures. In the case of the EU Refugee Crisis, social media was a tool that refugees used to develop their own solutions to the challenges of displacement—that is, navigating their way through the Balkan route and accessing services along the way—but such coping strategies also present risks, including vulnerability to surveillance and intimidation by extremist groups (Gillespie et al.,

2016). Ultimately, if community-orientated media initiatives are prevented from communicating about the matters that refugees deem relevant and vital, it is less likely that refugees will perceive them as a trusted source of useful information on which they can rely.

Further Research

Communication with Communities is a field of humanitarian response that is still finding and justifying its place within the broader scope of humanitarian intervention. While the provision of information may present protection risks to refugees, so too—in other cases—can the withholding of information. Further research on how and when protection logics are invoked as a justification for concealing information from refugees is needed. Such research could inform the development of guidelines for media initiatives that would assist practitioners in making decisions about the development of news content. Additionally, further research is needed on how media initiatives are perceived by the refugees they are intended to assist—and how they should be designed in order to generate perceptions of both relevance and trust.

References

Agier, M. (2011). *Managing the undesirables.* Cambridge: Polity.

Alencar, A., Kondova, K., & Ribbens, W. (2019). The smartphone as a lifeline: An exploration of refugees' use of mobile communication technologies during their flight. *Media, Culture & Society, 4*(6), 828–844. https://doi.org/10.1177/0163443718813486

Dekker, R., Engbersen, G., Klaver, J., & Vonk, H. (2018). Smart refugees: How Syrian asylum migrants use social media information in migration decision-making. *Social Media+Society, January–March*, 1–11. https://doi.org/10.1177/2056305118764439

Franklin, B. (2006). Preface. In B. Franklin (Ed.), *Local journalism and local media: Making the local news* (pp. xvii–xxii). London: Routledge.

Gillespie, M., Ampofo, L., Cheesman, M., Faith, B., Iliadou, E., Issa, A., …, Skleparis, D. (2016). Mapping refugee media journeys: Smartphones and social media networks. Research report. The Open University/France Médias Monde. Retrieved January 7, 2019, from http://www.open.ac.uk/ccig/research/projects/mapping-refugee-media-journeys.

Gillespie, M., Osseiran, S., & Chessman, M. (2018). Syrian refugees and the digital passage to Europe: Smartphone infrastructures and affordances. *Social*

Media and Society, January–March, 1–12. https://doi.org/10.1177/2056305118764440

Jack, V. (2017). 'There's no media for refugees': Information and communication in camps on the Thai-Burma border. *Pacific Journalism Review, 23*(2), 127–143. https://doi.org/10.24135/pjr.v23i2.93

Jansen, B. (2008). Between vulnerability and assertiveness: Negotiating resettlement in Kakuma refugee camp, Kenya. *African Affairs, 107*(429), 569–587. https://doi.org/10.1093/afraf/adn044

Kivikuru, U. (2013). Upstairs downstairs: Communication contradictions around two African refugee camps. *Journal of African Media Studies, 5*(1), 35–51. https://doi.org/10.1386/jams.5.1.35_1

Kutscher, N., & Kreß, L. (2018). The ambivalent potentials of social media use by unaccompanied minor refugees. *Social Media+Society, 4*(1), 1–10. https://doi.org/10.1177/20563051187644

Latonero, M., & Kift, P. (2018). On digital passages and borders: Refugees and the new infrastructure for movement and control. *Social Media+Society, 1*, 1–11. https://doi.org/10.1177/20563051187644

Robinson, L. (2004). *Communication patterns in refugee camps.* Paper presented at the annual International Studies Association conference, Montreal, Canada. Retrieved from http://citation.allacademic.com/meta/p74173_index.html.

Saltsman, A. (2010). Rumour versus information. *Forced Migration Review, 36*, 57–58.

Saltsman, A. (2011). Displaced Iraqis in Jordan: Formal and informal information flows, and migratory decisions in a context of uncertainty. *Refuge, 28*(1), 81–96.

Sommerfeldt, E. J. (2015). Disasters and information source repertoires: Information seeking and information sufficiency in post-earthquake Haiti. *Journal of Applied Communication Research, 43*(1), 1–22. https://doi.org/10.1080/00909882.2014.982682

United Nations High Commissioner for Refugees. (1999). *Protecting refugees: A field guide for NGOs.* Retrieved January 7, 2019, from http://www.unhcr.org/3bb9794e4.pdf.

United Nations High Commissioner for Refugees. (2006a). Operational protection in camps and settlements: A reference guide of good practices in the protection of refugees and other persons of concern. Retrieved January 7, 2019, from http://www.unhcr.org/448d6c122.html.

United Nations High Commissioner for Refugees. (2006b). Practical guide to the systematic use of standards and indicators in UNHCR operations. Retrieved January 7, 2019, from http://www.unhcr.org/40eaa9804.pdf.

United Nations High Commissioner for Refugees. (2011). *UNHCR resettlement handbook* (Rev. ed.). Retrieved January 7, 2019, from http://www.unhcr.org/46f7c0ee2.html.

Williams, G. (2006). Profits before product? In B. Franklin (Ed.), *Local journalism and local media: Making the local news* (pp. 83–92). London: Routledge.

CHAPTER 14

When Media Fuel the Crisis: Fighting Hate Speech and Communal Violence in Myanmar

Lisa Brooten

In 2012 and again in 2017, a surge of global media images brought to the world's attention the hundreds of thousands of destitute Rohingya Muslim refugees fleeing from attacks against them in Myanmar[1] and struggling to survive in overcrowded refugee camps in neighboring Bangladesh. These refugees recounted horrific tales of the violence spearheaded by Myanmar's extremist nationalist Buddhists against Rohingya Muslims that at this point has resulted in an exodus of more than 900,000 Rohingya Muslims, 745,000 of them after a major wave of attacks that began on August 25, 2017 (UNHCR, 2019). On that day, a group calling itself the Arakan Rohingya Salvation Army (ARSA) attacked a Myanmar security outpost, killing 12 military personnel and triggering a violent response from the Myanmar military.

[1] Prior to recent political changes in the country, the name Burma was generally used by those wishing to respect the opposition to the military regime, which changed the country's name in 1989 without a national referendum. Since this article deals with events since the reform process began, I use the now widely accepted name, Myanmar.

L. Brooten (✉)
Southern Illinois University, Carbondale, IL, USA

J. Matthews, E. Thorsen (eds.), *Media, Journalism and Disaster Communities*, https://doi.org/10.1007/978-3-030-33712-4_14

This crisis cannot be understood, however, without reference to Facebook, the as-yet unregulated transnational media giant which, while widely recognized as playing a role in *exacerbating* the violence, was also leveraged as part of a systematic campaign to create instability in the country and in this way helped to *create the crisis*. Many incendiary images were deliberately and systematically spread through the social networking platform by agents affiliated with the Myanmar military, promoting hate speech that triggered and fueled the violence and reinforcing historical prejudice against the Rohingya (Facebook, August 28, 2018; Mozur, 2018). The violent events in turn made for dramatic images and a high level of engagement by Facebook users.

This is especially dangerous in Myanmar, a country transitioning from a half century of military dictatorship, with Facebook becoming the primary source of information for many people, especially the estimated 21 million (approximately one third of the country) who are online and using the networking platform. Facebook itself is struggling to know how to respond to the myriad of problems caused by unanticipated uses of its platform in recent years. In its first major public response to the problems in Myanmar in August 2018, the company announced it was removing 20 Facebook and Instagram accounts found to be inciting violence toward the Rohingya, including those of Senior General Min Aung Hlaing and five other military officials.

Efforts to curb problematic content on Facebook are notoriously difficult. Reuters journalist Steve Stecklow (2018) identified more than 1000 examples of hate speech during one week in August 2018 in Facebook posts, comments, videos and images attacking Rohingya or other Myanmar Muslims. The material Stecklow (2018) found in his survey of Facebook posts violates Facebook's own rules, he writes, which "specifically prohibit attacking ethnic groups with 'violent or dehumanising speech.'" That same month, the UN Human Rights Council (2018, p. 14) released its *Report of the independent international fact-finding mission on Myanmar*, which found that in the case of the Rohingya Muslims, "the role of social media is significant. Facebook has been a useful instrument for those seeking to spread hate, in a context where, for most users, Facebook is the Internet. Although improved in recent months, the response of Facebook has been slow and ineffective."

The best hope for change and the focus of this chapter is the work of Myanmar's many active civil society organizations (CSOs) and their attempts to both counteract the immediate dangers and hold Facebook accountable for addressing these problems. I draw here on an analysis of the public communications between Facebook and Myanmar CSOs, and six in-depth interviews with five key local CSO members involved in the effort to counteract Facebook's dangers, one of them on two separate occasions. These change advocates must live with the on-the-ground consequences of Facebook's decisions to remove military accounts. These include not only the creation of new accounts and the resurfacing of problematic actors under other account names, but also sensitivity to the political environment for those CSOs working to hold Facebook accountable. In response to these concerns, I have agreed to maintain the confidentiality of my interviewees and so will not be using their names or organizational affiliations. The group of people in Myanmar actively addressing the concerns detailed here is small, making it difficult to maintain confidentiality if their organizations or their positions are identified. For this reason, interviewees are listed in the Appendix only by number and will be identified in the text with that number (I1 through I5) and the date of the interview.

This chapter examines the role played by Myanmar CSOs in raising concerns about Facebook and its influence over the country's public sphere. This case study of Facebook's use and misuse in Myanmar can help us to begin to identify the impact of social media in parts of the world with contexts significantly different from those in the United States and Europe, which demand the bulk of both academic and popular media attention. While social media have played a significant role in the violence perpetrated against the Rohingya community, technology does not function in a vacuum, and concerns about its use are meaningless without an understanding of context.

Military Dictatorship, Transition and Facebook's Rapid Rise

Prior to Myanmar's recent political reforms, the military, which took control in a 1962 coup, tightly controlled the media and other forms of expression. The military junta ruled for decades, using the country's challenges—including ethnic insurgencies and large-scale protests against military rule—to justify the need to "protect" the nation from disruptive forces. Their use of classic authoritarian strategies included tight control of the media, a network of state-run media to create divisive narratives and attempts to suppress civil society and people's capacity for critical thinking (Arendt, 1951; Fink, 2009). Despite its pariah status in the eyes of its own population and the world's democratic countries, the regime maintained a firm grip for nearly half a century.

When Myanmar's current transition process began is a matter of debate, but the top-down move for reforms resulted in problematic elections in 2010 and brought to power the interim quasi-civilian government of President Thein Sein, which began making significant changes. Elections in 2015 ushered in a civilian government under the leadership of the National League for Democracy (NLD), headed by Nobel Peace Prize Laureate and former political prisoner Aung San Suu Kyi. This moment held such promise—the possibility that genuine democratic reform could be more than a dream for the long-suffering Myanmar peoples. Yet the present NLD government must carefully navigate its relationship with the military, which retains enough power to retake control at any time by holding 25% of the seats in the country's parliament and controlling the key ministries of border affairs, home affairs and defense. While pre-publication censorship has been abolished and many problematic laws eliminated or revised, the government and military have also made repressive use of those laws that remain to quash dissent.

The liberalization of the telecommunications sector in 2013 rapidly decreased the price of SIM cards and exponentially increased people's access to 3G and 4G networks and affordable smartphones, most pre-loaded with Facebook. Suddenly, people in the rural areas gained a previously unprecedented degree of access to information and, unfortunately, misinformation. Over the course of only a few years, Facebook has reportedly grown to account for more than 85% of all internet traffic in Myanmar, where most people don't conduct internet searches or read web pages (Roache, 2018).[2]

Facebook in Myanmar

Facebook is ubiquitous in Myanmar. Most senior Myanmar government and military officials, government ministries, public figures, and civil society and business organizations have Facebook pages. Myanmar news outlets not only have Facebook pages but regularly cite the Facebook pages of government officials, ministries and other groups as sources in their stories. Even the fighting between the Myanmar military and minority groups is updated on both sides' Facebook pages, and this information in

[2] A recent media development report challenges the official figure for the percentage of Myanmar's population currently using Facebook, arguing that "the reach of the social networking site is far bigger than the official figure of 34 percent" (IMS-Fojo, 2018, p. 16).

turn is often cited in media. Facebook is now the dominant player in Myanmar's public sphere and a key site of struggle over the ongoing construction of the country's shifting, contested national identity.

Activists in Myanmar, as in many countries, have been warning about Facebook's size and structure, citing the company's lack of transparency and slow response to their concerns as dangerous messages spread rapidly, endangering local communities. Myanmar CSOs began raising concerns in 2012 about the spread of hate speech on Facebook and anticipating its role in the violence that was to follow. In early 2014, Myanmar ICT for Development Organization (MIDO), a local digital rights CSO, initiated the Panzagar ("Flower Speech") Campaign, to counter hate speech with positive messages. Myanmar CSOs point out how Facebook's design has been linked to many dangers, such as the easy spread of false, distorted and misleading information; the often-indistinguishable nature of different types of information with varying degrees of credibility; and the network's *prioritization* of strongly emotional messages, both positive and negative, which propel the content further (Vaidhyanathan, 2018). In 2015 tech entrepreneur David Madden warned Facebook executives about the misuse of their platform in Myanmar in a talk at the company's California headquarters (Stecklow, 2018). But it was not until after the violence in August 2017 that Facebook attracted more intense international scrutiny and began to publicly acknowledge and address concerns raised by CSOs in Myanmar.

2018 was a bad year for Facebook, even if we focus only on the platform's impact in Myanmar. In March, UN Special Rapporteur to Myanmar Yanghee Lee stated that Facebook had been used to incite violence against Rohingya Muslims (Lee, 2018). In April, Mark Zuckerberg appeared before the US House Energy and Commerce Committee to answer questions about Russian interference in the US elections. Shortly before this Senate appearance, Myanmar CSOs challenged Zuckerberg publicly with an open letter that called on the company to work more transparently with local stakeholders to address the dangers of Facebook in Myanmar ("Open Letter," 2018). Because these CSOs also pre-emptively contacted US Senators with evidence of Facebook's use to spread hate speech and incite violence, senators questioned Zuckerberg about the company's role in Myanmar and their plans to address the concerns raised by the CSOs. Then in late August, the UN Human Rights Council (2018) released the *Report of the independent international fact-finding mission on Myanmar*, which also found that Facebook had been used to spread hate. Shortly after this, Facebook announced its removal of the Facebook accounts of Senior General Min Aung Hlaing and other officials in Myanmar, including several named in the UN report.

In October 2018, the *New York Times* reported on the findings of its own investigation into Facebook in Myanmar, implicating the military in a systematic, coordinated and wide-ranging campaign using fake names and accounts. The *Times*' sources describe how military personnel "created troll accounts and news and celebrity pages on Facebook and then flooded them with incendiary comments and posts timed for peak viewership" (Mozur, 2018, para. 4). Facebook later confirmed that the campaign was systematic, covert and directly connected to the Myanmar military. Shortly after the publication of this report, Facebook announced it was taking down another series of military accounts (Facebook, 2018, October 15; Mozur, 2018). The *New York Times* report called these coordinated actions "among the first examples of an authoritarian government's using the social network against its own people" in an effort to "generate widespread feelings of vulnerability and fear that could be salved only by the military's protection" (Mozur, 2018, paras. 9 and 21).

State Responses

The National League for Democracy (NLD) and the military have responded similarly, denying any coordinated effort to drive out the Rohingya and arguing that international media report irresponsibly, exaggerating the situation and the hardships facing the Rohingya. They also accuse outsiders of focusing too much on human rights and not enough on national security. State Counselor Aung San Suu Kyi's Facebook page even claimed that Rohingya women were lying about being victims of sexual violence (Fisher, 2017). In December 2017, the NLD government made clear that UN Special Rapporteur to Myanmar, Yanghee Lee, who had openly condemned the human rights abuses she had heard about from Rohingya Muslim refugees, would no longer be welcome to visit Myanmar.

Despite expectations that the NLD government's 2015 election victory would improve the situation facing journalists and activists, under the NLD government at least 43 journalists had been arrested as of September 2018 (Human Rights Watch, 2019). The government has also continued its surveillance efforts. On August 22, 2017, just a few days before the August 25 ARSA attacks and the military's violent response, the Ministry of Transport and Communications submitted a plan to convene a Social

Media Monitoring Team, reportedly to track posts that may undermine "youths' moralities or peace and stability or security and rule of law on the open sources like facebook, messenger and twitter" (sic) (as cited in Budget spending, 2018, para. 6). They were given the go-ahead the following day, and the Team was formed in February 2018 (Budget spending, 2018).

Activists fear the surveillance powers of the Social Media Monitoring Team because counter-messaging campaigns intended to curb hate speech are often targets of online threats from Buddhist nationalists and others. Association with CSOs, engaging in media literacy and counter-messaging training could harm both these groups and the people they work with. As a consequence, the CSOs must keep a low profile by avoiding public association with specific online campaigns. This is especially a concern since Facebook began removing the accounts of nationalist Buddhists and government officials. Yet, these CSOs continue to work hard under the radar to address the very real dangers Facebook poses to Myanmar, its people and their efforts for democratic change.

Civil Society and Facebook

Myanmar's civil society organizations (CSOs) have from the start of the violence been among the most persistent voices raising concerns about Facebook's role in the devastating anti-Rohingya violence and have done so in an environment of widespread popular prejudice against the Rohingya. In early 2014 in an attempt to stem the hate speech and incitement of violence, a team of CSOs worked to translate Facebook's community standards into the Burmese language, in the (ultimately unrealized) hope that people would follow them (I2, August 7, 2018). The Panzagar ("Flower Speech") Campaign gained international attention in 2014 and at that time was the only campaign addressing hate speech in Myanmar (I5, February 8, 2019). Despite threats from the powerful nationalist Buddhist organization, Ma Ba Tha, the campaign and its key image of a flower quickly went viral, as people posted pictures of themselves with flowers in their mouths to symbolize peace and the principle of "right speech" or ethical conduct. Despite the campaign's popularity, the divisiveness and violence continued to spread throughout Myanmar, and CSOs' repeated warnings to Facebook that the Myanmar military was involved in coordinated efforts to use the platform as a weapon went unheeded (I1, August 7, 2018; Mozur, 2018; UNHRC, 2018).

Several CSOs are involved in this struggle, the most active of which are Phandeeyar and the Myanmar ICT for Development Organization (MIDO). Phandeeyar describes itself on its website as an innovation lab "spearheading the use of technology to accelerate change and development in Myanmar" through investing in tech start-ups and training to build the local pool of tech talent. MIDO focuses on the use of new technologies for peace, addressing issues of digital security, countering online hate speech and producing handbooks on mobilizing for peace and online security. Other key groups include Burma Monitor, a non-governmental organization (NGO) working to report on developments in the country's armed conflicts and on peace initiatives, and the Center for Social Integrity, an NGO described on its website as promoting a society "in which diverse peoples work collectively together towards peaceful coexistence and resilience" with a focus on peace and development through opening dialogues and training for communities and youth. Another is Equality Myanmar, which, according to their website, has the goal of "empowering the people of Myanmar through human rights education to engage in social transformation and promote a culture of human rights" primarily through advocacy campaigns, trainings and networking with other human rights advocates.

These CSOs conduct counter-messaging campaigns against hate speech, promote positive speech norms online and teach other groups how to create and run campaigns, especially Facebook campaigns, to amplify their efforts toward peace, tolerance, civil rights and digital rights. They also promote media literacy and monitor media for hate-filled posts and, as quickly as possible, counter and report them. For the last three years they have also organized a Myanmar Digital Rights Forum, held most recently in January 2019, which involves participants from government, civil society and media and tech companies, including Facebook. These CSOs are also working collectively on a broad digital rights campaign that moves beyond freedom of expression to also include issues of privacy, access to information, capacity building and other issues central to helping people take ownership of public discussion on key issues (I1, August 7, 2018). The impact of Facebook, however, remains a central concern for these groups, given the devastating humanitarian consequences of its misuse.

Conversations with Facebook

The 2017 ARSA attack on the Myanmar security outpost and the massive exodus of Rohingya that followed led to the first collective CSO efforts to pressure Facebook to change, with a jointly written briefing document and meeting with a Facebook team in December, their first visit to Myanmar after the August attacks. Despite this meeting, online hate speech and incitement to violence continued, so CSOs made the issue public in 2018.

> *Be warned and stay alert every time you go and eat. The Kalar*[3] *are planning to launch a Jihad on Monday 11 Sept. Warn your friends. The order to get ready with guns has already been issued in the army. Please forward this message in the chat box. Friends, brothers and sisters. let's unite.*
>
> —viral Facebook Messenger post, September 2017

> *Dear Islam brother, be warned and stay alert every time you go and eat. On 11 Sept in Yangon, Ma Ba Tha*[4] *and extremist nationalists will collaborate and they will launch an anti-kalar movement. Please forward this message to our brothers.*
>
> —viral Facebook Messenger post, September 2017

These messages, clearly coordinated to stir up unrest, were key triggers in the next collective CSO effort to pressure Facebook to address their concerns, an open letter to Mark Zuckerberg dated April 5, 2018 and signed by Phandeeyar, the Myanmar ICT for Development Organization (MIDO), Burma Monitor, the Center for Social Integrity, Equality Myanmar and the Myanmar Human Rights Educator Network. The letter responded to an interview Zuckerberg had conducted with *Vox*, published on April 2, in which he described Facebook's response to problems in Myanmar and referred to the two Facebook posts above:

> One Saturday morning, I got a phone call and we detected that people were trying to spread sensational messages through—it was Facebook Messenger in this case—to each side of the conflict, basically telling the Muslims, "Hey, there's about to be an uprising of the Buddhists, so make sure that you are armed and go to this place." And then the same thing on the other side.... In that case, our systems detect that that's going on. We stop those messages from getting through. (As cited in Klein, 2018, para. 80)

The CSOs' open letter disputed Zuckerberg's claims, providing a timeline that demonstrated the circulation of these messages for more than four

[3] "Kalar" is a derogatory term for dark-skinned people and is widely used to refer to the Rohingya.

[4] Ma Ba Tha was a high-profile group of extremist nationalist Buddhists, banned in May 2017 by government-appointed Buddhist authorities.

days before Facebook responded with a request for their source. It also described the impact of these messages:

> Far from being stopped, [these messages] spread in an unprecedented way, reaching country-wide and causing widespread fear and at least three violent incidents in the process. The fact that there was no bloodshed is testament to our community's resilience and to the wonderful work of peacebuilding and interfaith organisations. This resilience, however, is eroding daily as our community continues to be exposed to virulent hate speech and vicious rumours, which Facebook is still not adequately addressing. ("Open letter," 2018, para. 4)

The CSOs' open letter explained that Zuckerberg's reference to "our systems," which he claimed detected the problem, was in fact a reference to alerts forwarded to Facebook by the CSOs, rather than any "system" the company had put in place. As one interviewee put it, "their team had no idea [about the threat]—and as far as we could tell in terms of impact, the message went incredibly far" (I3, January 23, 2019). The letter also called for more proactive content moderation, greater engagement with local CSOs and investment in additional staff, especially native Burmese-speaking staff. It urged Facebook to be more transparent about hate speech reporting, the removal of abuse and fake accounts, the average review and response time, and review targets. This letter and its message were picked up and reported in international media, which raised awareness of the problem for people in Myanmar, where journalists had been reluctant to cover this story (I5, February 8, 2019). The open letter, as one interviewee put it, was "a turning point" (I4, January 23, 2019).

Facebook's Response

The CSOs' open letter, their lobbying of US Senators prior to Zuckerberg's testimony and the testimony itself had put Facebook in the spotlight, and as a consequence, the company scaled up its engagement in Myanmar. They sent a team to visit the country in June 2018. This time, they met with local civil society groups, programmers and hackers, who were able to provide their team with technical suggestions. This meeting gave CSOs the chance to engage on a broader level with Facebook, including with Facebook's tech and engineering staff, and to request changes to the platform (I1, August 7, 2018; I2, January 6, 2019; I4, January 23, 2019; I5, January 23, 2019).

Facebook's decision to remove Facebook accounts and pages linked to hate speech in August 2018 was the first time the company had removed the account of any country's military or political leaders (Dowling, 2019). A few months later, in October 2018, Facebook removed an additional 13 pages and 10 accounts linked to violence against Rohingya Muslims for "coordinated inauthentic behavior" (Nan Lwin, 2018, para. 12). The same reason was given again in December 2018 as Facebook purged nearly 600 additional accounts, pages and groups with links to Myanmar's military that were misrepresenting themselves as independent news, entertainment, beauty and lifestyle pages (Nan Lwin, 2018, paras. 1 and 5). Most recently, Facebook banned four ethnic "insurgent" groups: the Arakan Army, the Myanmar National Democratic Alliance Army, Kachin Independence Army and the Ta'ang National Liberation Army, which it described as "dangerous organizations" (Facebook, February 5, 2018). The wisdom of this latest move is questioned by CSOs, who were not consulted on this decision.

In its attempts to identify problematic content and hate speech, much of Facebook's efforts rely on artificial intelligence (AI), which is especially challenging, given the contextual nature of speech and its meaning (Milko & Hammond, 2018; I2 August 7, 2018; I3 & I4, January 23, 2019). As one Myanmar tech analyst familiar with the ongoing discussions with Facebook explained:

> They want to do machine learning on how to detect hate speech and specific kinds of behavior using IP addresses, how accounts behave and language algorithms... but in practice, hate speech is very subjective, and there's a question of freedom of speech. Also hate speech is very language specific—how much can AI handle this? (I2, January 6, 2019)

Activists argue that AI is not as cost effective as hiring local people in many countries in the global south, since labor costs are low and the local labor force are *uniquely* able to monitor content with the nuance necessary to prevent further incitement to violence (I3 & I4, January 23, 2019). Facebook management and staff outside of local contexts do not possess these skills, and even though local CSOs try to flag problems whenever they can, Facebook does not always act on these warnings. In just one of many examples, CSOs reported to Facebook a post from one frequent poster of hate speech that they believe called for violent action. The post called for people to come together and bring bamboo sticks, lauding the bravery of the Burmese and reminding readers that they had used bamboo sticks to triumph in their historic battles with neighboring Thailand. Facebook's response was that the post did not violate their community standards because it was not a direct call to attack and could promote important debate. Yet in the local context, the incitement was quite clear (I4, January 23, 2019). As another interviewee notes: "Facebook's understanding of our context is still very minimal so they always make mistakes. Sometimes they take out activists accounts. Sometimes they don't understand the seriousness of what we report and they say 'Oh this doesn't violate our community standards so we can't do that.' But it's serious, you know?" (I2, January 6, 2019).

The importance of local cultural capital is becoming increasingly apparent in attempts to prevent hate speech and the spread of misinformation on Facebook, a realization that has implications beyond Myanmar. Facebook in Myanmar, as elsewhere, relies on users to flag problematic content, yet they use an unsystematic strategy that has been strongly criticized (Dowling, 2019). By its own account, in 2015, Facebook had two people on staff who could speak Burmese. As of June 2018, that number had increased to more than 60, with a target of 100 by the end of 2018 (Facebook, 2018, August 15). Yet, even with a staff of 100 Burmese language speakers, it is a gargantuan task to weed out problematic speech from posts made by an estimated 21 million active Facebook users in Myanmar, especially when the staff are based in Facebook's offices in Malaysia or Dublin rather than Myanmar (Dowling, 2019; Stecklow, 2018; I3, January 23, 2019). One interviewee described what he sees as Facebook's misplaced self-confidence, saying "Facebook is like 'Oh, we can handle it,' but it's not that simple" (I2, January 6, 2019).

While it has taken a lot of time, effort and persistence by CSOs in Myanmar, Facebook has begun to listen to at least some of their advice (Dowling, 2019; I4, January 23, 2019 and February 8, 2019). Facebook has now hired a Myanmar public policy manager and is reportedly also recruiting a human rights director to assess how Facebook could be contributing to human rights abuses (Facebook, 2018, August 15; I3, January 23, 2019; I5, February 8, 2019). While CSOs are appreciative of the changes made so far, they remain deeply concerned about Facebook's power and the network's vulnerability to being leveraged once again to create division and incite violence in communities in Myanmar.

Counterbalancing Facebook

The horrific nature of the violence perpetrated on the Rohingya community was made possible by a process of dehumanization, a rhetoric of danger and an incitement to violence. The role Facebook played in spreading hate speech in Myanmar suggests it may be time for greater regulation of this global platform. Facebook "allows every user to post indiscriminately, develops algorithms that favor highly charged content, and is dependent on a self-service advertising system that precisely targets ads using massive surveillance and elaborate personal dossiers" and, therefore, as Vaidhyanathan (2018) argues, "cannot be reformed at the edges" (9). As one interviewee states,

> The meta-point here is that you cannot really have control over a public space when you are moderated by financial interests. If controversial content is what is going to drive your income then that's obviously going to be an issue. It's a business model question... Like the fact that they derive more profit from controversial content then they would from cat videos. (I3, January 23, 2019)

The pervasiveness of Facebook, especially in the world's poorest and most crisis-ridden countries, has significant implications beyond Myanmar. Among the most problematic assumptions of Facebook's founders were that the network's ability to connect people would bring equally positive results in all countries where it is introduced, and that a one-size-fits-all set of guidelines could be developed to help regulate content on this enormous platform (Fisher, 2018; Vaidhyanathan, 2018). It is clear from Facebook's experiences in Myanmar and the violence perpetrated toward the Rohingya community that the political, economic and social contexts within which a new technology is introduced determine how that technology is used, a reality that should now be starkly obvious to Facebook.

In contexts like Myanmar in which authoritarian powers can leverage Facebook to their own agenda, support for the civil society sector remains vital as a counterbalancing force. The civil society sector experiences firsthand the unintended consequences of a new technology and promotes indigenous and community-orientated solutions. The forms of regulatory control that can and should be developed are a conversation beginning to gain traction internationally as CSOs work to organize across borders. They argue that Facebook needs to reassess its impact in countries with very different political, judicial and economic structures than those in North America and Europe. "We see evidence of this across the global south—the system doesn't work and you cannot even apply their own rules properly," as one interviewee explained (I3, January 23, 2019).

The Myanmar CSOs working with Facebook argue that increasing investment in local partners should be a greater priority than technological "fixes."

> One of the things we've been telling Facebook is, if you really care about preventing problems, you shouldn't be testing your designs with focus groups of American users but you should be testing your stuff with people in countries like ours, so you can mitigate the worst. And if you want people with expertise about how the system can be misused, we probably have a lot more experience. (I3, January 23, 2019)

Local and community partnerships can help Facebook identify problematic actors and reduce their focus on content, which is significant given that much of the problematic content is circulated by specific actors with an agenda, who will be identifiable to local people with the necessary language skills and cultural capital. Moreover, local communities would also benefit from expanding their options for communication, including developing alternatives to platforms like Facebook. As one Myanmar tech analyst explained, "We really need to see beyond Facebook. We need to empower ourselves to not let one company be responsible for everything" (I2, January 6, 2019). This is a tall challenge in Myanmar, but perhaps an obvious next step in a country where Facebook's use is so ubiquitous and so clearly problematic in spreading hate speech and contributing to community division as has been witnessed in the ongoing Rohingya crisis.

Interviews

August 7, 2018—interview with I1 and I2
January 6, 2019—second interview with I2
January 23, 2019—interview with I3
January 23, 2019—interview with I4
February 8, 2019—interview with I5

References

Arendt, H. (1951/1973). *The origins of totalitarianism*. New York: Harcourt Brace Jovanovich.

Budget spending for Social Media Monitoring Team in line with policy. (2018, November 22). *Eleven Media Group*. Retrieved from https://elevenmyanmar.com/news/budget-spending-for-social-media-monitoring-team-in-line-with-policy.

Dowling, T. (2019, January 22). Shooting the (Facebook) Messenger Part II. *Tea Circle*. Retrieved from https://teacircleoxford.com/2019/01/22/shooting-the-facebook-messenger-part-ii/.

Facebook. (2018, August 15). *Update on Myanmar*. Retrieved from https://newsroom.fb.com/news/2018/08/update-on-myanmar/.

Facebook. (2018, August 28/October 15/December 18). Removing Myanmar military officials from Facebook. Retrieved from https://newsroom.fb.com/news/2018/08/removing-myanmar-officials/.

Fink, C. (2009). *Living silence in Burma: Surviving under military rule* (2nd ed.). New York: Zed Books.

Fisher, J. (2017, March 11). Hounded and ridiculed for complaining of rape. *BBC News*. Retrieved from https://www.bbc.com/news/magazine-39204086.

Fisher, M. (2018, December 27). Inside Facebook's secret rulebook for global political speech. *New York Times*. Retrieved from https://www.nytimes.com/2018/12/27/world/facebook-moderators.html.

Human Rights Watch. (2019). *Dashed Hopes: The criminalization of peaceful expression in Myanmar*. Retrieved from https://www.hrw.org/sites/default/files/report_pdf/myanmar0119_web3.pdf.

IMS-Fojo. (2018). *Myanmar's media from an audience perspective*. Myanmar: IMS-Fojo. Retrieved from IMS-Fojo 9.2018 Audiences-in-Myanmar_2018_finalweb.pdf.

Klein, E. (2018, April 2). Mark Zuckerberg on Facebook's hardest year, and what comes next. *Vox*. Retrieved from https://www.vox.com/2018/4/2/17185052/mark-zuckerberg-facebook-interview-fake-news-bots-cambridge.

Lee, Y. (2018). Statement by Ms. Yanghee Lee, Special Rapporteur on the situation of human rights in Myanmar at the 37th session of the Human Rights Council. Retrieved from https://www.ohchr.org/EN/NewsEvents/Pages/DisplayNews.aspx?NewsID=22806&LangID=E.

Lwin, N. (2018, December 19). Facebook shuts hundreds more pages linked to Myanmar military. *The Irrawaddy*. Retrieved from https://www.irrawaddy.com/news/facebook-shuts-hundreds-pages-linked-myanmar-military.html.

Milko, V., & Hammond, C. (2018, May 16). Facebook reveals AI weaknesses on hate speech. *Frontier Myanmar*. Retrieved from https://frontiermyanmar.net/en/facebook-reveals-ai-weaknesses-on-hate-speech.

Mozur, P. (2018, October 15). A genocide incited on Facebook, with posts from Myanmar's military. *New York Times*. Retrieved from https://www.nytimes.com/2018/10/15/technology/myanmar-facebook-genocide.html.

Open Letter to Mark Zuckerberg from Myanmar CSOs. (2018, April 2). Retrieved from https://drive.google.com/file/d/1Rs02G96Y9w5dpX0Vf1LjWp6B9mp32VY-/view.

Roache, B. (2018, January 16). The way people use the internet in Myanmar—And, in particular, the dominance of Facebook—Has implications for businesses and advertising agencies. *Frontier Myanmar*. Retrieved from https://frontiermyanmar.net/en/what-myanmars-facebook-supremacy-means-for-business.

Stecklow, S. (2018, August 15). Hatebook: Why Facebook is losing the war on hate speech in Myanmar. *Reuters*. Retrieved from https://www.reuters.com/investigates/special-report/myanmar-facebook-hate/.

UN High Commissioner for Refugees (UNHCR). (2019, February 15). Joint UNHCR IOM Press Release. United Nations seeks US$920 million for Rohingya humanitarian crisis in 2019. *UNHCR/IOM*. Retrieved from https://www.unhcr.org/en-us/news/press/2019/2/5c667ecf4/united-nations-seeks-us920-million-rohingya-humanitarian-crisis-2019.html.

UN Human Rights Council (2018). Report of the independent international fact-finding mission on Myanmar. United Nations General Assembly A/HRC/39/64. Retrieved from https://documents-dds-ny.un.org/doc/UNDOC/GEN/G18/274/54/PDF/G1827454.pdf?OpenElement.

Vaidhyanathan, S. (2018). *Antisocial media: How Facebook disconnects us and undermines democracy*. New York: Oxford University Press.

CHAPTER 15

Afterword: The Shifting Domain of Disaster Journalism

Mervi Pantti

Media, Journalism and Disaster Communities highlights the importance of re-addressing the relationship between media and disasters when the nature of both disasters and media is changing. The different case studies demonstrate that disasters are not just news events, but in addition to impacting communities and societies, they also shape and leave traces on local media ecology. Questions about how international and national news media frame the significance and meaning of disasters and what kind of challenges disasters pose for their news reporting have been at the core of journalism studies. But questions about how local media responds to and is shaped by disasters have been less often visited. Clearly, the social role of local media after a disaster cannot be deduced from studies on national or international media.

The analytical focus in this volume is through the concept of disaster communities as articulated by Matthews and Thorsen in the 'Introduction'—communities that are affected by and recovering from the impacts of disaster. Through engagement with different empirical case

M. Pantti (✉)
University of Helsinki, Helsinki, Finland
e-mail: mervi.pantti@helsinki.fi

J. Matthews, E. Thorsen (eds.), *Media, Journalism and Disaster Communities*, https://doi.org/10.1007/978-3-030-33712-4_15

studies, the book traces the different forms of media and journalism produced by and for disaster communities and how such media perceive and accomplish their missions to offer alternative perspectives to national or international media, to give voice to those affected by disasters and to support relief, rebuilding and recovery. The premise of the book is that local news media are deeply embedded in the dynamics of disaster communities and significantly affect a community's response to a disaster. Indeed, literature on local media has shown that local journalism is pivotal to a sense of community, civic communication and public participation (e.g., Nielsen, 2015). Undoubtedly, disasters are situations that especially highlight these cultural, social and political functions of local media. Importantly, 'local media' are not defined here only by a geographical nearness to people but by an emotional attachment and connection to a place they symbolically construct and maintain.

In taking up these goals, the contributions of *Media, Journalism and Disaster Communities* seek to accomplish three important tasks: first, to redress the current imbalance in favour of studies on international and national news coverage through a focus on local and community media; second, to widen the dominant attention to text analysis to include studies on the reception and production of disaster news; and third, in so doing, to capture the complexities of the encounters between local media and those affected by disasters at this juncture. This 'Afterword' examines the contributions of this collection and offers reflections on some of the key issues in the field of mediated disaster studies that connect the chapters: namely, locality, listening and accountability.

Local and Community Media in a Networked News Ecosystem

Local and community media outlets are changing as part of a wider transformation of the media environment affecting national and international media. Local media ecosystems have become hybrid, and there are important forms of interdependence between national and local media, between 'new' and 'old' media, as well as between legacy news media and other information providers such as non-governmental organisations (NGO). Despite the doom and gloom regarding the decline of local journalism in the current digital news environment, changes in the news ecosystem have also allowed new types of (hyper)local media and grassroots community media to emerge, both online and in print. In short, digital technologies have reshaped the local media ecology by diversifying both the tools available for disseminating information and the actors involved in news production (e.g. Powers, 2018). On the other hand, local media is also facing serious challenges, as the business models on which commercial local and regional news media are based are no longer sustainable. While this collection is not directly concerned with the economic and organisational challenges and transformations of local media, it nevertheless advances our understanding of the relationship between local media and disasters. In addressing the question about how disasters shape the local media ecology, and vice versa, the case studies point to the need to reconsider the role of place in the media-disaster nexus. Research has shown that the increasingly networked systems of global communication and the concomitant spatial and temporal reorganisation of social life have had an important role in globalising disasters (Cottle, 2014). At the same time, local or community journalism is increasingly understood as serving communities of interest rather than those defined solely by a specific location.

One of the key messages of the collection is that a sense of community needs to be determinedly constructed, and the local media potentially plays a key role in this symbolic work. Jamie Matthews' chapter—'Kesennuma's *Building for the Future* and Ishinomaki's *Rolling Press*'—exemplifies the fluid nature of today's disaster communities and the role of community media in maintaining 'local' social networks. The chapter also addresses the essential but so far little studied question about what motivates community news producers to create local content (Williams & Harte, 2016). By examining two grassroots media initiatives that were launched in the aftermath of the 2011 Japan disaster, the chapter shows that in a contemporary disaster context, new forms of community media emerge as a response to the formations of globally dispersed disaster communities. The producers are motivated by their ability to maintain connections between local people and people interested in their plight and to represent the affected community and its ongoing recovery after the departure of international and national reporters. In this case, the non-profit media initiatives provided English-language local news specifically for international volunteers who travelled to the Tōhoku region to assist in disaster relief and remained concerned about the recovery of affected communities in that area.

Further, the collection importantly reminds us that people are excluded from media coverage if they are not understood as belonging to a particular community. Forced displacement, within or outside of national borders, is one of the world's most intractable humanitarian crises. Travelling migrants problematise a fixed sense of locality and signal the changing meaning of 'disaster communities'. Refugees at temporary stopping points, as Victoria Jack's chapter—'Informing Refugee Communities in Greece'—illustrates, form a particularly vulnerable community, which is exacerbated by the deprivation of reliable information and exposure to rumours and misinformation in social media. Jack addresses the special information needs of the forced migrants in camp environments by focusing on the role of community media in safeguarding refugees' right to information. Here, 'disaster community' is first and foremost formed by shared experiences and information needs rather than by a connection to a geographic territory.

While displaced people cannot be served by the local or national media, digital tools make it possible for various advocacy groups, increasingly engaged with communication efforts, to address specific target groups directly. Thus, community media may represent an alternative, non-profit media which serves communities ignored or unrecognised by commercial local or national media. This is the case with the News that Moves project that Jack examined; it focuses on identifying misinformation and providing practical news designed to meet the constantly changing needs of refugees on the move in Europe.

While *Media, Journalism and Disaster Communities* makes a valuable contribution by problematising and redefining the spatial dimensions of local media, it also reminds us that we need to take into account the differences in practices and quality of local media across different national and local contexts. To be sure, local media does not necessarily fulfil the ideal of benefitting and engaging disaster-affected communities for various reasons, including lack of professionalism, economically precarious conditions, political control, commercialisation and media concentration. As the analysis of the news consumption practices of the Donbas community in Ukraine shows (Chap. 8), local media is not perceived as having any community value to the community nor as providing relevant and reliable information—beyond the TV guide, as noted by a focus group participant.

Listening to the Disaster Communities

Another key theme running throughout the collection emphasises the media's obligation to listen to those affected by disasters. Journalism scholarship has paid a considerable amount of attention to voice, which is considered necessary for democracy and social justice but is also denied to many citizens (Couldry, 2010). The concept of listening shifts the focus from marginal voices to those who have the power to 'shape who and what can be heard in media', as Tanja Dreher (2010, p. 85) writes. Mathew Charles' chapter—'Changing the Story of Urban Violence in El Salvador'—shows that listening to marginalised groups, including the voices of perpetrators of violence, can lead to a greater social cohesion in post-conflict societies. As he writes, the online news outlet *El Faro* urged their readers to listen to voiceless groups and, by taking their stories seriously, departed from mainstream media coverage. The paradox noted by scholars is that while journalists now have more tools to engage with audiences, they have less time to listen to marginal voices (Wasserman, 2018). Such practices of active and empathic listening merit more attention in studies on local journalism.

From this perspective, local or community media are understood as having privileged knowledge about and access to affected communities. Thus, the potential advantages of local media are clear in terms of working closely with and understanding disaster-affected communities. Moreover, audience research shows that people tend to expect local news media, above all, to care about and listen to the community (Nielsen, 2015). In this collection, several contributions examine and compare how international, national and local media are, or are not, listening to people affected by disasters. The stakes here are high; if local media do not allow affected people to speak, it is less likely that their concerns will be recognised by national audiences and political elites. Moreover, the result of not representing the interests of a community or listening to a community is the erosion of trust in local media.

In the context of disasters, listening is connected to the understanding of disasters. The ways in which disasters are framed and narrated inevitably shape how they are responded to and how the needs of the destroyed communities are accounted for. Scholars have pointed out that even declaring an event a 'disaster' is not without political and social implications, since such a declaration performatively shapes the way in which these events are responded to by the media, governments and humanitarian NGOs (Calhoun, 2013). In accordance with current theorisations, the volume underlines the importance of seeing disasters not as sudden events but as processes with sociopolitical causes and long-lasting implications. Mainstream media coverage, however, has been predominantly event-focused: disasters typically fade from view within a matter of days or weeks after the emergency phase (except when they happen to 'us'), allowing those who are not caught up in them to imagine them as extraordinary and contained events, after which life goes back to normal. Thus, while disasters may receive wide international and national coverage, attention to rebuilding community may be scarce. However, when international reporters go home, local journalists remain in the disaster area and may take an active role in rebuilding and in supporting the emotional recovery of devastated communities and keeping their concerns on the agenda (Matthews, 2017).

Chindu Sreedharan's and Einar Thorsen's chapter illustrates that journalists working within or belonging to affected communities have a deep continuing connection with and commitment to the affected community, even if it is to the detriment of their journalistic professionalism and objectivity. That the Nepalese 'survivor-journalists' expressed antipathy for foreign reporters who they saw as lacking in empathy and understanding of local experiences and selling suffering as a commodity to their audiences clearly illustrates how local journalists with pre-existing bonds with the community delineate their social role as distinct from that of national or international 'parachute reporters'. Inevitably, unusual conditions also shape journalists' professional norms and inform their practices; this, in particular, questions journalists' roles as undetached recorders of events (e.g., Usher, 2009). As this chapter shows, surrendering norms of impartiality and detachment may enable journalists to better listen to their community.

Scholars have repeatedly shown that the proximity of a media outlet to a news event influences journalistic practices. Contributions in this book show that there are clear differences in how local and national news media listen to affected communities. However, as already stated, we should not assume that local media serves communities equally in different contexts. Anna Roosvall, Matthew Tegelberg and Florencia Enghel's chapter—'Media and Climate Migration'—investigates how climate migration from three US islands is framed in local and transnational reporting. The chapter, as well as Kyle J. Holody's chapter on the news coverage of the Parkland mass shootings, shows that local areas and affected communities are covered in greater depth by local media than by transnational media. While local media attempt to listen to the affected indigenous communities and recognise their political agency, transnational journalism globalises the fates of islanders in an attempt to create ideological links between those events and the climate change potentially affecting us all.

Such findings correspond to the idea of local journalism as representing communities to themselves and the outside world, offering information directly relevant to the communities, fostering cohesion, providing a space where a community feels listened to and advocating for the community. As local media markets are shrinking in many countries, such roles cannot be taken for granted. During the last decade, as summed up by Dave Harte, Andy Williams and Jerome Turner (2017), local news has become 'less plural; less local in its orientation; less embedded in, and reflective of, local communities; as well as less critical and independent from its (overwhelmingly elite) sources'. While some communities have access to a wide variety of local news through newspapers, radio and hyperlocal websites, in others, local journalism is deteriorating and being replaced by news that appeals to broader audiences (Napoli, Stonbely, McCollough, & Renninger, 2017).

This collection shows that disaster communities' needs for reliable information and meaningful interpretations are not met equally. The question is not only about the disappearance of traditional local news outlets but also about the quality of the journalism being produced and distributed within communities: even if local news media exists, it does not necessarily provide a public arena for a community's concerns and critiques. As Paola Prado and Juliet Pinto show in their chapter about the devastating Samarco disaster in the state of Minas Gerais in Brazil, commercial local news media did not offer an alternative, community-focused perspective to national news outlets but echoed the voices of official sources in national media that framed the collapse of the dam holding iron mining waste mainly in terms of economic losses.

Local Disaster Coverage and Social Accountability

Mediated disasters are moral events, as they provide encounters with vulnerabilities, suffering, injustices and wrongdoing. They are also political events, in which journalism can play an important role in democratic accountability, placing questions of responsibility on the table. After the immediate response phase, media plays an important role in the identification of causes—such as failed or absent policies—and in the facilitation of potential solutions which may then influence policymaking. In sociological disaster studies, inequalities and injustices unveiled by the media have been seen as an important catalyst of political awareness and can prompt popular demands for policy solutions that would prevent or prepare for future disasters (e.g., Birkland, 2006).

Another key theme in the book revolves around the question of the news media's role in assigning responsibility and promoting social and political change after a disaster. The various roles the news media plays in social accountability—from exposing potential failures to providing disaster-affected populations with opportunities to voice criticism of public authorities—are an increasingly important field of research. It needs attention because digital technologies have transformed the ways in which networked media can take advantage of the new methods of distribution to publicly scrutinise the conduct of authorities, mobilise counter-narratives and make the voices of ordinary people count.

Kurt Barling's chapter, 'Is Local Journalism Failing?—Local Voices in the Aftermath of the Grenfell and Lakanal Fire Disasters', explicitly addresses the impact of local journalism on accountability processes. Did local journalism fail to fulfil its political role to expose authorities' failures? Here, we see that listening to the communities forms a vital part of the media-initiated accountability process. In the case of the Grenfell disaster, local media did not cover the concerns of local residents about fire safety and, thus, failed to offer an adequate opportunity to put pressure on local authorities. More generally, the decline of professional journalism at the local level generates a distance between media and the communities it reports on, which may lead to a diminished capacity to call authorities to account. Barling claims that by failing to listen to the community, local journalism left 'an accountability vacuum', which contributed to the Grenfell tower fire disaster. Arguably, an 'accountability vacuum' likely also contributed to another dam collapsing and contaminating the environment with mining debris in Minas Gerais a few years ago (as studied by Prado and Pinto, see Chap. 2).

In the current media environment, focusing solely on the role of legacy media as the enablers of social accountability processes through investigative watchdog journalism is a limited approach. New digital media platforms have created a new dynamic and new hopes for informal accountability processes. Importantly, the relationship between the news media and networked publics may strengthen both these groups' capacities to hold the powerful responsible for their failures; for example, journalistic exposés of apparent wrongdoing may lead to online appeals and demands for officials to explain their actions. Moreover, new participatory media technologies enable citizens to voice their critiques and demands directly to public authorities and private organisations.

Lisa Brooten's chapter—'When Media Fuel the Crisis: Fighting Hate Speech and Communal Violence in Myanmar'—illustrates the weaponising of digital platforms for accountability by looking at how local civil society organisations attempted to hold Facebook accountable for its role in the proliferation of hate speech and incitement of violence. The paradox is that while digital platforms offer powerful new tools for civil society actors and journalists to scrutinise the conduct of authorities, they also foster actors and content that are not motivated by liberal and democratic norms of accountability. In Myanmar, Facebook was used to exploit prejudice against the Rohingya and promote hate speech that fuelled violence against the minority group. Such disinformation and hate campaigns present a considerable departure from the role of social media platforms as instruments of democratic accountability. From the point of view of media accountability, the question is whether Facebook and other social media platforms are just 'channels' or if they can be held accountable for the hate speech shared on their platforms. Brooten argues that in authoritarian contexts such as Myanmar, it is vital for companies such as Facebook to work closely with local people and NGOs to identify problematic actors that seek to create community division.

Final Remarks

It is difficult to do justice to the richness of the contributions of *Media, Journalism and Disaster Communities*—the first collection that emphasises the importance of local media and disaster community approach in the study of the current media-disaster nexus. Disaster studies have emphasised the global reach of disasters in an increasingly interconnected world. Similarly, media and journalism research has focused on media representations and journalistic practices in a global or national context. Yet the global nature of contemporary disasters does not mean that 'local' no longer matters. Given that people's vulnerability and adaptation to disasters is rooted in their everyday material condition, it is important to understand disasters from the perspectives of disaster communities. Local media, understood as diverse forms of media tied to specific communities, ideally play an important role in producing social cohesion and solidarity after disaster. The same key questions about building community, listening and responsibility traverse more or less explicitly through the collection. A widespread belief holds that community-based media are powerful because they care about the affected people and help to build social capital after a disaster. There is truth in this viewpoint even in the face of declining local media. While we learn that in some contexts the relations between traditional local news media and communities are not characterised by distinct closeness and trust, in the current fluid media environment, new forms of locally relevant media, new mediated ties, are developed within disaster communities.

References

Birkland, T. A. (2006). *Lessons of disaster: Policy change after catastrophic events.* Washington, DC: Georgetown University Press.

Calhoun, C. (2013). The idea of emergency: Humanitarian action and global (dis) order. In D. Fassin & M. Pandolfi (Eds.), *Contemporary states of emergency* (pp. 29–58). New York: Zone Books.

Cottle, S. (2014). Rethinking media and disasters in a global age: What's changed and why it matters. *Media War & Conflict, 7*(1), 3–22.

Couldry, N. (2010). *Why voice matters.* London and Thousand Oaks, CA: Sage.

Dreher, T. (2010). Speaking up or being heard? Community media interventions and the politics of listening. *Media, Culture & Society, 32*(1), 85–103.

Harte, D., Williams, A., & Turner, J. (2017). Reciprocity and the hyperlocal journalist. *Journalism Practice, 11*(2–3), 160–176.

Matthews, J. (2017). The role of a local newspaper after disaster: An intrinsic case study of Ishinomaki, Japan. *Asian Journal of Communication, 27*(5), 464–479.

Napoli, P. M., Stonbely, S., McCollough, K., & Renninger, B. (2017). Local journalism and the information needs of local communities. *Journalism Practice, 11*(4), 373–395.

Nielsen, R. K. (2015). Introduction: The uncertain future of local journalism. In R. K. Nielsen (Ed.), *Local journalism: The decline of newspapers and the rise of digital media* (pp. 1–25). London: I. B. Tauris.

Powers, M. (2018). *NGOs as newsmakers: The changing landscape of international news.* Columbia University Press.

Usher, N. (2009). Recover from disaster. *Journalism Practice, 3*(2), 216–232.

Wasserman, H. (2018). Relevance, resistance, resilience: Journalism's challenges in a global world. *Journalism, 20*(1), 229–232.

Williams, A., & Harte, D. (2016). Hyperlocal news. In T. Witschge, C. W. Anderson, D. Domingo, & A. Hermida (Eds.), *The SAGE handbook of digital journalism* (pp. 280–293). London: Sage.

Index[1]

[1] Note: Page numbers followed by 'n' refer to notes.

J. Matthews, E. Thorsen (eds.), *Media, Journalism and Disaster Communities*, https://doi.org/10.1007/978-3-030-33712-4

The manufacturer's authorised representative in the EU is Springer Nature Customer Service Centre GmbH, Europaplatz 3, 69115 Heidelberg, Germany. If you have any concerns regarding our products, please contact ProductSafety@springernature.com

Printed and bound by CPI Group (UK) Ltd, Croydon, CR0 4YY
30/09/2026
02225461-0006